SECOND EDITION

Designing
SURVEYS

To Beth and Cookie

SECOND EDITION

Designing
SURVEYS
A Guide to Decisions and Procedures

Ronald Czaja
North Carolina State University

Johnny Blair
Abt Associates Inc.

PINE FORGE PRESS
An Imprint of Sage Publications, Inc.
Thousand Oaks • London • New Delhi

For information:

Pine Forge Press
A Sage Publications Company
2455 Teller Road
Thousand Oaks, California 91320
(805) 499–4224
E-mail: sales@pfp.sagepub.com

Sage Publications Ltd
1 Oliver's Yard
55 City Road
London EC1Y 1SP
United Kingdom

Sage Publications India Pvt. Ltd.
B-42, Panchsheel Enclave
Post Box 4109
New Delhi 110 017 India

Library of Congress Cataloging-in-Publication Data

Printed in the United States of America

Czaja, Ronald.
Designing surveys : a guide to decisions and procedures /
Ronald Czaja, Johnny Blair.—2nd ed.
    p. cm.
Includes bibliographical references and index.
ISBN 0-7619-2745-X (cloth) — ISBN 0-7619-2746-8 (pbk.)
    1. Surveys. 2. Sampling (Statistics) I. Blair, Johnny. II. Title.
HA31.2.C93 2005
001.4′33—dc22

2004014521

05   06   07   08   09   10   9   8   7   6   5   4   3   2   1

Acquiring Editor:	Jerry Westby
Editorial Assistant:	Vonessa Vondera
Production Editor:	Sanford Robinson
Copy Editor:	Richard H. Adin
Typesetter:	C&M Digitals (P) Ltd.
Cover Designer:	Glenn Vogel

Contents

Preface

In the preface to the first edition, we said that one purpose in writing this book was to inform readers that a great deal about surveys is still unknown and to encourage researchers to conduct methodological studies as part of their substantive investigations. In the ensuing decade, new technologies have impacted how surveys are conducted, but that basic point remains. The most prominent new technologies affecting survey research include the World Wide Web; cellular telephones; pagers; answering machines; caller ID; call forwarding and blocking; the national Do Not Call Registry; portable telephone numbers; fax machines; high-capacity, high-speed laptop computers; and personal digital assistants (PDAs). Thus, survey design and implementation continues to be very much a work in progress.

The new technologies have, in the main, aided survey research, but also created new and difficult problems and obstacles. On the positive side, laptop computers and Internet surveys have given us more versatility, better data quality, and faster survey results following data collection. On the negative side, telephone sampling frames are less complete and efficient, it takes more effort to contact respondents while refusal rates have increased, and Internet surveys allow us to collect large amounts of data in a short period of time, but the population coverage and response rates are often suspect. Internet coverage, number portability, and cell phones have created challenges that still need to be met.

New technologies have helped both novice and experienced survey researchers. The Internet has great potential to make conducting a survey economically feasible for the beginner with limited resources. Large organizations, such as the Census Bureau, have taken advantage of PDAs and wireless technology to improve the efficiency of in-person data collection. But it is important to remember that the principles of scientific survey design remain unchanged. Careful probability samples must still be selected. Frames and data collection modes must still provide adequate population coverage. Questions must still be written to elicit accurate, unbiased answers. And response rates are still the main indicator of survey reliability. Ethical issues in the treatment of human subjects have not gone away; in fact, the public's sensitivity to matters of privacy and the use of data may have been raised by the impact of more powerful technologies for data linkage and data dissemination. This new edition does not replace or change advice provided earlier, but builds on it.

Several people provided assistance in the revision of this book. We wish to thank Peg Brant and Kevin Stainback for their reactions and helpful comments to selected chapters. We appreciate the timely and extensive efforts of Ed Blair who gave us comments on the entire manuscript. Bruce Cheek and Judy Cline provided outstanding secretarial and manuscript preparation assistance.

Two people worked with us for more than a year on this revision and provided invaluable assistance. Elizabeth Eastman gave extensive substantive and editorial comments on numerous drafts. Alix Rosenberg, our undergraduate assistant, helped on many tasks including literature searches and census information, gave substantive comments on chapter drafts and did the indexing.

Finally, we thank our editor at Sage, Jerry Westby, and other support staff—Vonessa Vondera, Sanford Robinson, and Richard Adin—for making the final tasks of putting a book together go as smoothly and trouble-free as possible.

As always, we appreciate any comments or suggestions readers have about this edition. We can be reached at ronc@sa.ncsu.edu and Johnny_Blair@abtassoc.com

Acknowledgments

The authors and Pine Forge Press gratefully acknowledge the contributions of the following reviewers:

Audie L. Blevins
University of Wyoming

Harry L. Wilson
Roanoke College

Margaret Fitch-Hauser
Auburn University

Suzanne Wilson
*University of Illinois
at Urbana–Champaign*

G. David Johnson
University of South Alabama

Edward W. Wolfe
University of Florida

Frederick J. Kviz
University of Illinois at Chicago

Rita Kirk Whillock
Southern Methodist University

1

An Introduction to Surveys and to This Book

It would be difficult to name another social science method that has so quickly and pervasively penetrated our society as the sample survey. In less than two generations, the notion of relying on relatively small probability samples to measure, among other things, public attitudes and behaviors has grown from a little-noted curiosity to the dominant practice. It is commonplace to look first to surveys to gauge such important national issues as pre-election preference for presidential candidates, consumer confidence in the economy, the level of unemployment, the accuracy of the decennial census, the rates of occurrence of different types of crimes, or the public's knowledge about health issues such as acquired immunodeficiency syndrome (AIDS) risk behavior. In the marketplace, surveys are used to test public reactions before new products are introduced, and to evaluate customer satisfaction with goods and services. In academia, surveys provide data for social scientists testing models and hypotheses in fields as diverse as economics, sociology, and psychology. In fact, so successful has been the survey method in so many aspects of research and public policy that one must be cautious not to use a survey in inappropriate circumstances.

This science on which we place such impressive burdens is, if not exactly a toddler, no more than an adolescent, strong in some ways, inexperienced and unsure in others. In some of its various aspects, the sample survey may rely on powerful mathematical tools; in other aspects, it is dependent on empirical experience with incomplete theoretical underpinnings; and in still other aspects, it makes use of conventions and practices that only recently rose above the level of professional folk wisdom.

If this state of affairs is perplexing to the professional practitioners of this young science, it is especially daunting for the novice or the occasional user who needs to conduct a survey or interpret a survey's results.

The Practice of Survey Research

Survey research is inherently interdisciplinary. Sampling and estimation procedures require a knowledge of probability theory and statistics. Data collection involves persuasion of respondents and then, on some level, social interaction between them and interviewers. Interviews and questionnaires depend on cognition, recall, language comprehension, and discourse. The use of laptop computers and Internet questionnaires requires programming skills in various software. Although few professional practitioners are expert in statistical theory, psycholinguistics or sociolinguistics, and cognitive psychology, each of these disciplines speaks to some area of survey research. When one adds to these the management expertise necessary to conduct even moderate-size surveys, the enterprise seems daunting. Yet surveys, even very large and complex ones, are done all the time. And although, to the extent we can determine, they often produce useful, reliable, and valid results, in many cases they do not.

When the survey enterprise is successful, it is largely because at each stage in designing and conducting surveys, researchers focus on a relatively small number of key scientific principles and practical guidelines that are applied in a series of key decisions. This book provides practical guidance for the reader to do the same. Although it will not produce the high-level expert described above, this book is a realistic guide to the effective conducting of small-to-moderate-scale surveys.

Even modest-size surveys typically require considerable time, material, money, and assistance. One can find examples of surveys designed and implemented by the lone researcher, but they are the exception. Unlike some scientific or scholarly enterprises, surveys are usually a team effort of many people with diverse skills. Even if the researcher who formulates the research questions also designs the questionnaire and analyzes the data, that person will almost always use help in collecting the data and entering it into a computer file, as well as in performing numerous clerical tasks. Thus, whether the survey is an organization's endeavor or a class project, there is a division of labor, a coordination of tasks, and the incurrence of time and costs. Accomplishing all this with limited resources and, at the same time, maximizing the survey's quality, requires numerous decisions by the researcher.

A project guided by such a series of judicious decisions will make the best use of all its resources—not just money, materials, and labor, but also time, information, and talent. It is on these decision points, which often require compromises or trade-offs, that this book focuses. We show how the researcher, after the specification of the research questions, reconciles the ideal with the possible through all aspects of survey and questionnaire design, sampling, and data collection. To illustrate the process, we have selected actual research projects whose design and implementation issues complement each other. Each chapter highlights the decision points in these research projects and draws on other examples as well.

Surveys are not appropriate in all circumstances. Therefore, before discussing some general aspects of sample surveys, we offer the following guidelines for deciding whether or not to conduct a survey in the first place. After all, if we try to use a survey in inappropriate situations, we can hardly expect success.

Surveys are based on the desire to collect information (usually by questionnaire) from a sample of respondents from a well-defined population. The questionnaire, alternatively referred to as the instrument, typically contains a series of related questions for the respondents to answer. The questions are most often, but not always, in a closed format in which a set of response alternatives is specified. The resulting numerical, or quantitative, data are then entered into a data file for statistical analysis.

This thumbnail description of a survey implies several simple, but essential, conditions for the appropriate use of surveys. The target population must be clearly defined, usually in terms of a simple combination of demographic characteristics and geographic boundaries. For example, a target population might be all persons age 18 years or older residing in households with telephones in Maryland. The population should be defined so that its members can be unequivocally identified.

In addition, we must be convinced that the majority of respondents will know the information we ask them to provide. It makes little sense to ask people questions, such as the net worth of their families, that many in the targeted population, maybe most, will not be able to answer.

Finally, the goals of the analysis should be to answer the research questions, test hypotheses, estimate population characteristics, model a set of variables, or other well-defined goals using the appropriate statistical procedures.

The Uses of Surveys

Making use of sophisticated statistical methods such as multiple or logistic regression, hierarchical analysis, and analysis of variance, researchers can use survey data to test hypotheses and study causal relationships between variables. The pairing of survey data with advanced analytical methods has become, along with theory building and psychology experiments in laboratories, one of the foremost means of social investigation. But on the most basic level, the idea of a survey begins with the desire to know, which is to say measure, some unknown characteristic of a population. That characteristic is often a simple parameter, such as the proportion of the adult U.S. population who smoke cigarettes or own a DVD player, the total number of households that employ domestic help, the proportion and characteristics of households that have access to the Internet, or the average annual household expenditure for entertainment (however we choose to define that vague term). More complex statistics can be measured as well; for example, the ratio of savings or investments to income. A survey also may be aimed at only one segment of the population, such as the elderly (those age 70 years and older), households with young children (younger than age 5 years), or people who own computers. It is also common for researchers to be interested in analysis of a particular subgroup in addition to the total population results. So, for example, we may want to determine the general population's perceptions of crime, as well as analyze perceptions by racial group, or by city dwellers compared to suburban residents. On the other hand, we may not be interested in the general population at all, but only in farmers' attitudes about the environmental effects of pesticides or in college students' average school-year expenses.

Much of survey methodology, particularly sampling theory, is aimed at obtaining estimates of such population parameters in a rigorous fashion. Most of the focus of this book is the same. Yet, it is important to understand that these survey methods are equally applicable to the needs of the model builder and the describer of populations. That is, there is not one survey methodology for the hypothesis tester and another for the policy analyst seeking only to measure parameters such as those noted above. In fact, a distinguishing property of surveys is that the resulting data are often put to both descriptive and analytic uses. Government surveys, for example, tell us things like the proportion of persons with a regular place to seek health care (National Health Interview Survey) or the total number of adults who were unemployed last month (Current Population Survey). These same data are used by sociologists, economists, and others with interests much different from obtaining those simple point estimates.

Overview of the Survey Process

Although the focus of this book is on survey design and implementation, we must begin with the process of specifying the study's analysis goals. We should recognize that this step is, in fact, usually a *process* that develops in the course of the survey's development rather than a full-blown plan from the start. We should, at the outset, have an analysis plan that answers certain questions, such as: Are we interested in obtaining only point estimates or in more complex measures as well? Which variables are key to our research? Which are secondary? Are we concerned with particular subgroup estimates? Exactly what analysis is planned? What will our data tables look like? If we have multiple goals, what are their priorities? Exactly what are the research questions we hope to answer? The answers to these questions will govern the design of the survey. It is also likely that some components of the analysis plan cannot be exactly specified or will change as we develop the questionnaire. During instrument development, we are forced to think much more concretely about how particular questions will be used and what—exactly—they mean. We will come up against limits on questionnaire length and have to decide which questions are retained and which are dropped. So, while lip service is often paid to the need for a fully specified analysis plan early on that remains unchanged throughout the study, the reality is often quite different.

Regardless of the subsequent uses of the data, the survey researcher must always carefully define the target population about which information is needed; obtain or develop a sampling frame that lists that population; decide on a sample design that describes precisely how population members will be selected; develop an **estimation** plan[1] for computing values of the population parameters from the sample estimates; decide on a data collection method; and specify detailed data collection procedures to ensure the quality of the data in each data collection stage.

Each of these steps involves making choices between alternative methods, and each of these technical decisions is *inherently* also a decision about the expenditure of resources. This book provides guidance in making decisions in the context of fixed resources. These resources will sometimes be financial, but more often (especially for the first-time researcher working with colleagues who volunteer their time) will have to do with time or other nonmonetary support. Unlike a large government or business survey, in which the goal is to achieve a certain level of precision while minimizing cost, our survey model is a fixed-cost model, meaning that our goal is to obtain the best measures we can within a given "budget."

Implicit in this decision-points approach is the notion that, while there are many wrong ways to go about designing and conducting a survey, there is no one right way. Instead, there are different approaches suited to particular goals, target populations, resources, and other constraints. We hold that (for the types of goals noted above) **probability sampling,** giving every population member a known (nonzero) chance of inclusion, is always preferred over convenience sampling, which takes whatever people are most easily obtained. In choosing the survey respondents with probability samples, we have a statistical basis for making statements, based on the sample, about the population. Still, the choice of available probability sample designs is wide and the consequences of that choice important. Moreover, each design requires a series of further choices about specific implementation procedures.

We try to write questionnaires that will convey to respondents a uniformly understood request for information. This task is a difficult one for which we have fewer scientific principles to follow than we do for sample design. We know what we mean or want a question to convey and are often baffled as to why others don't quickly know it too. As soon as we begin testing our draft instrument, unintended meanings begin to emerge. There are many ways of testing a questionnaire, which we shall describe in detail, but a large part of the benefit of testing seems to come from simply exposing the questionnaire to people not involved in its construction. Whether they are potential respondents, friends and colleagues, or survey experts, they do not know the exact intent of the question and so must depend on what it literally says, on its words and context, to infer that intent.

Researchers can easily underestimate the impact of context while writing individual questions, but it becomes pivotal when those items are combined in the instrument and each question may exert a subtle influence on the interpretation of those that follow. As Caron (1992, p. 150) points out, ". . . the *ambiguity* (author's italics) inherent in all natural languages invests the context of the utterance with a determining role in the construction of meanings." Nevertheless, our efforts to avoid ambiguity must not oversimplify or undermine a question's substantive intent.

In data collection, we will always be concerned with selecting a method appropriate to the kinds of questions we intend to ask as well as with obtaining cooperation from as many respondents as possible. But those concerns alone will not tell us whether mail, Internet, telephone, group administration, personal visit, or some combination of these is best for our study; nor, once that choice is made, will they guide us through the various subsequent procedures to carry out the data collection. It is in such fundamental decision making and procedural tactics that this book offers guidance.

Throughout, we also address concerns about the effects of choices between, for example, reduced list (sampling frame) coverage versus cost of the frame; the collection of more data from a small number of respondents versus less data from a larger number; bigger samples versus better cooperation rates; or inclusion of sensitive questions versus the risk of more respondent refusals to answer them. At the same time, issues of production and quality control during data collection draw considerably on our attention and resources.

One final principle underlying our approach is the counsel of Sudman (1976) to keep in mind how good a survey needs to be for the purposes at hand. While as survey researchers we strive to conduct the highest-quality project feasible, we also recognize that all surveys do not need the same levels of precision and reliability. A survey conducted by a faculty member under a grant will likely have quality imperatives that the undergraduate class project will not; and neither requires the levels of precision of a major government survey such as the Current Population Survey (CPS) or the National Health Interview Survey (NHIS). So while we aim for excellence in every phase of the project, we must be realistic as to what is both necessary and achievable. That realization, while not easily quantifiable, will help guide many of our choices—and may even determine whether we decide to go ahead with the survey.

Once we have decided that a survey is both appropriate for our research and within our means, we begin to specify its major requirements. First, what is the time frame and budget? If the study is part of a 1- or 2-semester class, at what point do the data need to be available? If the study is for an outside sponsor, are there time constraints posed by the sponsor's needs? Once the question of project schedule is answered, we are faced with a series of decisions about the survey's characteristics, including the necessary sample size and approximate interview length. Before we can begin designing the survey in detail, we need at least a general notion of these study conditions, and we need to determine whether they are in line with our resources.

Thus, we cannot begin too early to list the resources at our disposal and balance them against the aims of our research. In this way, we continue to shape the broad outlines of our project and prepare to make the first major decision confronting us, the selection of a data collection method.

A Brief Summary of This Book

Before we discuss data collection, we describe the general stages in the development and completion of a survey in Chapter 2: survey design and preliminary planning; making sampling decisions; pretesting; final survey design and planning; data collection; data coding, data-file construction, and analysis. This chapter emphasizes the critical points in the survey process and the many decisions that must be made. The chapter concludes with an example of a time schedule for a national telephone survey. The major tasks and the time allocated for each are outlined.

Chapter 3 briefly describes mail, Internet, telephone, and face-to-face surveys and points out the advantages and disadvantages of each method in terms of required resources, questionnaire design, and data quality. The data collection methods are compared with regard to costs, time needed for data collection, limitations on questionnaire length and complexity, types of questions that can be asked, use of visual aids, establishment of rapport, sensitive questions, and comments on response rates. A number of examples from actual studies illustrate these points in addition to a few examples of combinations of methods.

Chapter 4, which addresses questionnaire construction, begins with a discussion of the questionnaire design process. This process approach provides both a framework for the discussion and a list of steps that the new researcher can follow. Next is a simple set of characteristics that constitute a good questionnaire. Then, through numerous examples, we show how to write individual questions that balance the information desired with the ability of respondents to provide valid and reliable answers. We also provide guidelines for deciding which questions to include in the instrument. Chapter 5 continues questionnaire design with a discussion of typical questionnaires. The crucial issue of how to begin the questionnaire is given even more attention than previously in recognition of the growing concerns about survey response and the potential for serious levels of nonresponse bias. The reader is shown how to organize the questionnaire from its introduction, through each section, to its conclusion. Chapter 6 is a comprehensive treatment of questionnaire testing. Drawing on the most recent research in instrument design, we show readers how to use each of the major pretesting methods and how to interpret the test results. This chapter includes an expanded section on behavior coding and updates the discussion of cognitive interviewing to include recent research that has implications for pretesting practices.

Chapter 7 introduces sampling basics, beginning with the differences between nonprobability and probability samples. The chapter focuses on the probability sample and illustrates why it is the preferred method. Particular attention is given to the key tasks in developing and selecting probability samples, including defining the population, constructing and evaluating sample lists, and handling unexpected situations and common problems. The chapter also responds to one of the most frequently asked questions, "What sample size do I need for my survey?" Hypothesis testing and power are included in this discussion. A number of examples illustrate how to use census data in planning a survey and for estimating whether the number of interviews with important subgroups will be adequate for the analysis planned.

Chapter 8 provides a number of sampling examples. We describe many of the decisions that must be made in selecting a directory-based community telephone sample and how to select a list-assisted and a national random-digit telephone sample. We discuss procedures for randomly selecting a respondent within a sample household for interview. The chapter concludes with two examples of how to select student samples from lists.

Chapter 9 is a comprehensive treatment of nonsampling error in survey data collection. The first half of the chapter explains the fundamentals of bias and variance as they pertain to data collection, particularly the impact of unit and item nonresponse. This section ends with a summary of the main measures of survey quality. The second half of the chapter is devoted to procedures for reducing nonsampling error in both interviewer-administered and self-administered surveys.

Chapter 10 covers ethical issues in surveys, survey budgeting and a detailed guide to preparing a methodology report, an essential aspect of a carefully executed survey often omitted from introductory textbooks. The reader is shown the importance of a complete description of the survey methods. Quality profiles are defined and their major components described. Then the reader is shown, through many examples, both how to decide what issues to include in a particular study's report and what to say about them. The chapter ends with a guide to the major recent literature on survey methodology.

Note

1. Boldface terms in the text are defined in the Glossary/Index.

2

Stages of a Survey

The five general stages in the development and completion of a survey, as shown in Exhibit 2.1, are

1. Survey design and preliminary planning

2. Pretesting

3. Final survey design and planning

4. Data collection

5. Data coding, data-file construction, analysis, and final report

This chapter discusses the decisions called for at each of the five stages. The following are especially critical points in the design and implementation of a survey (see boldface boxes in Exhibit 2.1):

- *Design survey:* Key preliminary decisions about method of data collection and sampling are made.
- *Pretest:* Decisions from multiple activities are tested and evaluated.
- *Revise survey design and operations plan:* Final design decisions are made based on the pretesting activities.
- *Collect data:* Data collection and quality-control procedures are carried out.

Stage 1: Survey Design and Preliminary Planning

Stage 1 involves the issues discussed in Chapter 1: specification of the research problem and the research questions that the survey will address. At this stage

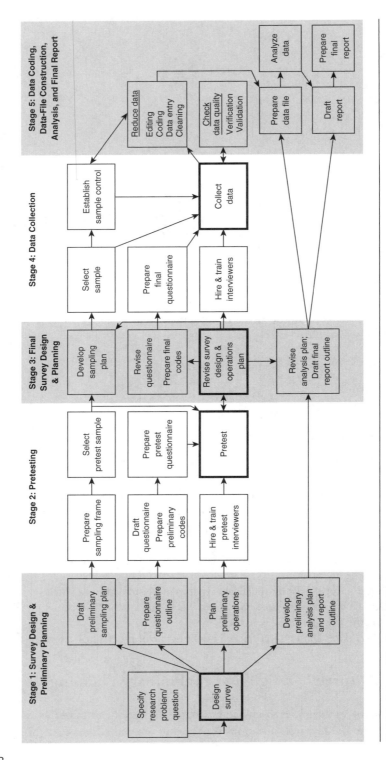

Exhibit 2.1   The Stages of a Survey

12

we must decide the goals of the research and determine how best to accomplish them within the available time and resources. We first need to specify the goals of the survey.

- Is the goal to test a hypothesis? For example,

  Females are more likely than males to believe that a woman should be allowed to have an abortion for any reason.

- Is the goal to test a causal model that suggests a series of interrelated hypotheses? For example,

  1. People with conventional values are less likely to have ever used amphetamines;

  2. People with drug-using friends are more likely to have used amphetamines; and

  3. People with conventional values are less likely to have drug-using friends.

- Is the goal to estimate the proportion of people who hold a certain belief or that engage in a specific behavior? For example,

  What proportion of the population believe that our criminal justice system is working well?

  What proportion of the population were the victims of robbery in the last year?

- Is the goal to study specific topics or a group of people over time to see what changes occur in attitudes or behaviors, or, to see if we can design interventions to modify behaviors? For example,

  The General Social Survey has interviewed adults for the past three decades on topics such as abortion, religious service attendance, attitudes toward the death penalty, self-rated social class, neighborhood safety and many other subjects.

  The National Longitudinal Survey of Youth began interviewing more than 12,000 youth in 1979. These individuals have been interviewed every 1 to 2 years about number of jobs, earnings and other labor market activities.

  The Community Intervention Trial for Smoking Cessation was a 4-year intervention involving 22 communities. The goals were to get smokers to quit and to maintain abstinence, especially for heavy smokers (≥ 25 cigarettes per day).

Many of the above goals would typically require general population studies of adults. Other goals may not concern the general population but might relate to specific **subgroups** in the population. For example, we may be interested in knowing what dental hygienists think about changing the requirements for state licensure, what the attitudes of university faculty are about joining a union, or what proportion of a migrant worker's income is spent on health care. Another possibility is that our goal may not be concerned with individuals at all, but with organizations, groups, businesses, or governmental units. We may want to know what the annual membership rates are for specific types of organizations, or the dropout rate for high school juniors and seniors for every high school or school district in a state. Specifying the goals provides the framework for the next set of decisions: Who or what is the population of interest? What should be the method of data collection?

Making Sampling Decisions

One of the most basic sampling questions must be answered in stage 1: Who or what is the population of interest? In studies of individuals, three considerations are the eligible age of the respondents; the geographic area to which we want to generalize the results; and whether we include people who live in households, group quarters, the homeless, or some combination of these. For example, if we are doing a general population survey, should we include people age 18 years and older? Or age 21 years and older? Or should we include 17-year-olds? When making this decision, we need to keep the research questions in mind. If we are conducting a survey on the public's perceptions of crime on public transportation, we would want to include both respondents who use public transportation and those who do not. We should consider setting as the minimum age of eligible respondents that at which people begin regularly using the system, perhaps age 16 years or even 14 years, if we are conducting the research in areas where students use public transportation to get to and from school.

When studying groups, we must decide who in the group or organization is the most knowledgeable individual and can provide accurate information. For example, if we are interested in learning the average cost and length of a hospital stay for a specific type of surgery, we would want to survey the chief financial officer at each hospital in our sample. Sometimes determining the best person may require a phone call to the site or organization. We might also contact the head of the organization and request that he or she or a knowledgeable delegate complete the interview or the questionnaire. If we need different types of information (e.g., usual cost for cataract surgery, number of births, number of cancer patients seen) from each group, information might have to be collected from two or three different individuals within the organization. The key point is that it is important to specify or locate knowledgeable respondents—also known as *informants*—when seeking information about a group.

Another consideration is the geographic area we want the survey to represent. Let's continue with the example of a survey of public perceptions of crime on public transportation. In some cities, public transportation serves not only the central city but also some suburbs. However, coverage in the suburbs is usually not as extensive as it is in the central city, and is often provided by a different company. We must decide how inclusive the survey should be. Should we focus only on the central city? On the entire transportation system? On the central city and certain suburbs? There is no magic formula or rule of thumb for making that decision; it must be based on the articulation of the research questions and the available resources.

When these decisions are made, trade-offs must be considered. Consulting census data will help in determining the ramifications of expanding or contracting the age definition of eligible respondents. As we discuss later, the more interviews we conduct with a given group or subgroup, other things being equal, the more confidence we can place in the results. Assume, for example, that we can afford 500 telephone interviews. If we use an eligible age definition of 21 years and older, all 500 interviews will be conducted with people in this age group. However, if we decide to expand the definition to people age 14 years and older, some interviews will be conducted with people age 14 to 20 years, and fewer than 500 interviews will be conducted with people age 21 years and older. If we learn from census data that the 14- to 20-year-old age group comprises 13% of the eligible population and our survey is conducted properly, then approximately 65 interviews (13% of 500 = 65) will be conducted with respondents age 14 to 20 years, and 65 fewer interviews will be conducted with respondents age 21 years and older. Thus, while the confidence in our overall survey results does not change, the confidence in the results for respondents age 21 years and older does change because we have fewer interviews with them. Similar trade-offs must be considered in the decision to include or exclude suburban respondents in the study universe. The logic is the same. If we have a fixed budget and decide to include some or all of the suburbs, then some of the city interviews would have to be allocated to the suburbs, resulting in less confidence in the city results. (This issue of trade-offs is discussed further in Chapter 7.)

Similar decisions must be made in the study of organizations, institutions, or businesses. For example, if we are interested in studying school systems, we must decide if we want to include both public and private schools. If hospitals are the population of interest, we must decide whether we are interested in only for-profit hospitals or whether we should include nonprofit hospitals.

In addition, we may want to survey only hospitals with a minimum bed size or hospitals that provide specific services. We resolve these issues by examining the goals of our research and by assessing our available resources. These are decisions that you, the principal investigator, must make. Usually there are no right and wrong answers to the decisions that must be made. You must assess your priorities and balance them against your resources and time schedule.

A second sampling question that must be addressed is the availability of a sampling frame from which to select a random sample from the study population. The **sampling frame** is the source (or sources) that includes the population of eligible people or groups. Sometimes the frame matches perfectly the population we want to study. The list or resource includes, for example, every individual or school in the population of interest. On other occasions, the frame may include more than the population of interest, and we would need to screen sample members to determine whether or not they are eligible. Or the frame may not include all of the eligible population members, and then we must decide whether it is adequate for our purpose. Here are two examples of sampling frames:

1. If we want to conduct a survey of all students enrolled at a university, we would ask the registrar's office for a list of all currently enrolled students that includes their addresses and telephone numbers. This list would be our sampling frame.

2. If we wanted to do a telephone survey of adults in North Carolina, we might obtain copies of telephone directories that include, together, all areas of the state. These books would constitute our sampling frame.[1]

Telephone directories are among the most common sampling frames for general-population surveys. However, using a telephone directory raises an important question: How well does the sampling frame represent the eligible population? In other words, what proportion of the eligible population does the frame include? What kinds of people may not be included on the frame? Obviously, people without telephones are excluded, as well as those with unlisted telephone numbers and most people that have only a cell phone.

Sampling frames frequently do not include everyone in the eligible population. When the population definition and the sampling frame do not match, we need to ask the following questions: What percent of the population is missing? Who are these people? If we exclude them, how will the results for the **dependent variable**—the variable we want to explain or estimate—be affected? We are asking, in essence, how much bias we will allow in our survey. There are three major sources of bias with telephone frames: unlisted residential telephone numbers, nontelephone households, and cell-phone-only subscribers. In most urban areas of the country, the percentage of households with an unlisted number is much larger than the percentage of households without telephones. Fortunately, **random-digit dialing (RDD)** techniques, which we discuss in Chapter 8, allow researchers to overcome some of the bias of using a telephone directory that misses unlisted residential numbers. However, the bias from excluding nontelephone households cannot be overcome without assuming the costly task of finding them. The percentage estimates of nontelephone households varies by the source consulted. The 2000 Census estimates that 2.4% of occupied households (U.S. Census Bureau, 2000a, Table DP-4) have no telephone service, while the March 2003 Current Population Survey (CPS) estimates nontelephone coverage at 4.5% (Belinfante, 2004).[2] Additionally, about 7.5 million people have gone totally wireless and do not have a landline household telephone (Carroll, 2003). The estimates are that 3% to 5% of cell phone users have given up their landlines and these folks are young, urban, mobile, and professional. Thus, it is very likely that approximately 6% of households nationally do not have a landline telephone. Therefore, on average, we would need to visit in person about 17 households ($16.6 \times 6 \approx 100\%$) to find one without a telephone or a landline connection—a great expenditure of time and money.[3] (Keep in mind that these are national figures and that the proportion for a particular area is likely to be different.)

Most researchers who conduct surveys ignore the bias that results from excluding nontelephone households. They do so for three reasons. First, they assume the nontelephone households will have a small or negligible effect on the final results. Because the telephone group is so much larger than the nontelephone group, the differences between them would have to be great before the results from the nontelephone group could affect the final results. Second, the cost to include nontelephone households is too great, especially because they constitute so small a percentage of the total population. Third, total sample results are of primary importance to the researcher, and subgroup results are not of major importance. In some cases, however, these assumptions are not valid, and the impact of excluding nontelephone households must be considered carefully.

The proportion of households without telephones varies by region of the country and by demographic characteristics. In March 2002, in the Northeast, 3% of households did not have a telephone; the percentages in other regions were the Midwest, 4.2%; South, 5.5%; and West, 2.9% (U.S. Census Bureau, 2003). Income is strongly correlated with telephone coverage. For households below poverty, the percentages by region were 9.6%, 15.1%, 16.0%, and 7.5%, respectively. Eleven states had less than 85% coverage for households with incomes below $10,000 annual income in 1984 dollars (approximately $17,954 in 2003): Alabama, 81.1%; Arkansas, 82.7%; Illinois, 80.0%; Kentucky, 84.7%; Louisiana, 82.8%; Mississippi, 82.4%; Nevada, 84.7%; New Mexico, 81.7%; Oklahoma, 82.0%; Virginia, 84.9%; and Wyoming, 84.4% (Belinfante, 2004). There are differences by owner (1.2%) versus renter (6.6%) for nonphone ownership; for those below poverty, the percentages increase to 5.9% and 10.0%, respectively. Among demographic groups, the highest incidence of nontelephone households is found in American Indian and Alaskan Native households, 11.9% (U.S. Census Bureau, 2000b). The percentage of households with telephones increases as education, family income, and age increase. Those age 65 years and older have the highest telephone coverage.

We need to ask whether it is worth trying to include nontelephone households in our survey. The answer, again, depends on the research problem, our resources, and the location of our research. If we are interested in the opinions of the unemployed or those living below the poverty index, we will want to consider a combination of methods or a method other than a telephone survey, which would exclude as of 2000, 1 of every 6 unemployed persons and 1 of every 4 households below the poverty index. (We should keep in mind that these are national ratios and that the numbers for any specific community may differ significantly.) We also know that nontelephone households have lower incomes or are households in transition.[4] People in these types of households may hold different opinions or engage in different behaviors from those of people in households with telephones, depending on the variables under study. We should always consult census data or other available data about our survey area to help in planning the research (see Chapters 7 and 8).

Designing the Questionnaire

The types of information we need to collect from respondents and how best to elicit that information are two key decisions that must be made early in the survey design phase. We need to know whether we will be asking many **open-ended questions** (e.g., "What is the most difficult problem facing families today?"), mostly **closed-ended questions** (e.g., "Do you approve or disapprove of abortion?"), or both. In the former category of questions, the respondents answer in their own words. In the latter category, respondents choose from a list of provided responses. We should also consider whether we will be asking attitude, knowledge, or behavior questions, and determine what types of demographic information we need because these decisions can affect our choice of a data collection method. A personal interview or face-to-face survey is probably best if it is essential to (a) hand respondents lists of choices from which they are to select an answer, (b) give them other types of visual aids to help formulate answers (i.e., pictures that we want them to evaluate, response cards to be categorized, etc.), or (c) have them consult personal records or perform other memory-assisting tasks. Similarly, if we need to ask a number of open-ended questions, we would not conduct a mail survey because many respondents will not complete a questionnaire that requires a lot of writing. We always want to make the respondent's task as easy as possible, thus minimizing the reasons for not responding.

Determining Available Money and Time

Two other important factors to consider in the preliminary design stage are the amounts of money and time we have for conducting the survey. Money and time determine how many interviews we can afford; whether we can hire interviewers to collect the data and coders to code it, and how many of each; and, in some cases, how large a geographic area we can include in the survey. Web surveys are the fastest and cheapest, followed by mail and then telephone surveys; face-to-face surveys are the most expensive. When it is possible to conduct a group-administered survey—for example, with students in classrooms—this may be as cheap and quick as a Web survey. In terms of time, telephoning is faster than face-to-face interviewing; however, the time to complete both is somewhat dependent on the size or geographic distribution of the sample. Web and mail surveys are less affected by these variables, because the procedures for conducting them are the same regardless of sample size and geographic distribution. For mail, however, the time schedule is fixed and usually not affected by the sample size. A first mailing is followed by a reminder postcard in approximately 2 weeks; a few weeks later, a second cover letter and questionnaire are sent; and a few weeks later, a third mailing or a telephone call is made. Thus, the data collection phase of a mail survey usually lasts 8 to 10 weeks, regardless of the sample size and its geographic distribution.

Analyzing Data and Reporting Survey Results

In the preliminary design stage, we must give some thought to the kinds of analyses we need to answer the research questions and the time it will take to prepare a study report, papers, or other products proposed from the research effort. If we, ourselves, are not doing all of the end-stage work (i.e., data analysis and interpretation, writing, and typing), we must have some idea of how the tasks will be shared and how much time to allocate to them. This step is important not only for estimating the total cost of the survey, but for planning the number of persons and person-hours necessary to complete the work. Time schedules and costs can vary widely, depending on the type and the scope of the survey work.

Stage 2: Pretesting

In stage 2 of the survey process, we begin testing our initial design decisions. This step entails preparing the sampling frame, record-keeping forms, and survey questions, and then testing these items to see how well the process is working. To begin, we assemble a sampling frame or, better yet, determine if an acceptable one already exists for our target population.

Drafting the Questionnaire

When putting together initial drafts of our questionnaire, borrowing questions from other research studies is acceptable, even encouraged. While we are reviewing past studies with the same or similar research problems, we take notes on how the researchers defined their concepts and the questions they used to measure these concepts. When we agree with what they did, we are free to use the same wording, unless they copyrighted the questions or scales.[5] If we agree with parts of what they did, we use the parts we agree with and then add our own ideas and wording. Using someone else's questions has another advantage: By asking the same questions, we can compare the results in our survey area with the results from the previous research.

There are two caveats regarding questionnaire development to keep in mind. First, we should never assume that because we are using someone else's questions and they reported no problems with them that these questions will work as well with our study respondents. No matter how many questions we borrow from others, we must always pretest the questionnaire before we start the data collection. Pretesting, like a dress rehearsal before opening night, is one of the most important components of a survey. In fact, if we are developing or asking many new questions, we may want to plan two or three pretests to make sure that the respondents understand the questions and that what we are trying to achieve is working at least on a miniature scale. It is better to keep doing pretest interviews until we feel comfortable with the questionnaire—which may require 30, 40, or 50 interviews—than to start the main data collection, find problems, and either stop the data collection to fix them (which is nearly impossible) or end up collecting interviews that have obvious measurement problems.

The second caveat concerns the number of questionnaire drafts. Expect to do many drafts of the questionnaire to get to the stage where all the problems seem to be worked out and the respondents' interpretations are incorporated into the questions. Often the difference between a good study and a poor one is that in the former, researchers look for or anticipate problems, and in the latter, researchers assume that if the questions are answered, the data are valid. We should guard against this faulty assumption.

Using Interviewers

For telephone and personal interview surveys, we need to start thinking about how many interviewers we will need and begin assembling and writing training materials that explain how we want things done. Both written training materials and the training of interviewers are necessary whether professional interviewers, volunteers, or students are doing the interviewing.

Selecting and hiring interviewers must be closely coordinated with sample selection. For personal interview surveys, we want the interviewers to live reasonably close to the selected households to minimize the amount of travel time and expense required to contact respondents. In a class project, for example, to interview respondents on the north side of town, we would assign students who live close by. Because telephone interviewers usually conduct interviews at a central, supervised location, where they live is unimportant.

The amount of time allocated for pretesting determines the number of pretest interviewers that we will need. If we have a week for pretesting and we need 30 personal interviews, we would probably want to hire and train 6 interviewers, each of whom would conduct 5 interviews. For telephone interviewing we might want to use three interviewers, each of whom would average ten completed interviews. We use our best interviewers, if possible, because we want critical feedback on any potential problems with our questionnaire or survey procedures.

Debriefing Interviewers

At the end of each pretest, all interviewers meet with researchers in a debriefing session to go over each question in the questionnaire and all the survey procedures to determine where the problems are and to propose possible solutions. The group meeting encourages an interactive exchange of ideas in which one person's comments triggers a comment or idea someone else may have forgotten. Feedback from the meeting, along with the reading of the actual interviews, provides the basis for revising the questionnaire.

Deciding on Pretest Methods

Pretesting can be done in a number of different ways and steps. In the early phases, the purpose of pretesting is to get feedback on individual questionnaire items. In the later phases, the purpose is to test the entire questionnaire and the survey procedures. Questionnaire items are usually tested both informally and formally. As we begin writing questions, we should think about how respondents will react to or interpret specific questions. If appropriate population members are available, you can get a quick reaction by trying the questions out on them. Colleagues and students may also be asked to try out questions and provide feedback. The first draft of the questionnaire may be tested informally on family, friends, other students, and so forth. When the questions seem to be working well in these informal trials, it is time for a more formal test of the questionnaire with real respondents in the survey area.

Two procedures are now used more and more in the development of questionnaires. One is cognitive interviewing and the other is focus groups.[6] Cognitive interviews are one-on-one sessions, with an interviewer and a respondent. Interviewers ask respondents what thoughts occurred to them as the questions were read to them. In addition, the interviewer often probes the respondent to ask about the meaning of a particular word or to ask them to paraphrase the question, as a way to assess their understanding of it. For example, the questionnaire might contain the following question topics:

- What does *food safety* mean to you?
- How important are food safety issues to you personally?
- What does *biotechnology* mean to you?
- What are your impressions of using genetic engineering or biotechnology to change the foods we eat?

As part of the pretest, we want to find out what the terms *biotechnology* and *food safety* mean to respondents and how they arrive at the answers to the survey questions. Understanding the cognitive processes that respondents use in answering questions will help us to write better questions and collect better data.

Another procedure is the use of **focus groups**, small groups of people assembled to discuss topics, issues, or other matters that will help us in writing questions or conducting the survey more efficiently. There are two important ways that focus groups are used. First, prior to instrument development, the groups can be used to ensure that important issues are included, or otherwise learn more about the topic of the survey. For example, in a survey about low-income housing, the researchers certainly will know something about the topic, but can easily overlook other issues that are less widely known or that have recently emerged. Second, the focus group can be a tool to test a draft questionnaire, particularly one designed for self-administration. In this instance, group participants can be asked to complete the questionnaire on their own, and then discuss it as a group. Groups typically comprise eight to ten people with similar characteristics: all men, all women, young black males, working-class people, and so forth (Greenbaum, 1998). The composition of the groups is partly determined by the subject matter of the survey. The groups should be constructed so that the members feel comfortable when talking. If the survey is about sexual behaviors related to risk of AIDS infection, it would be important to have separate groups for men and women. On the other hand, for a survey about racial issues and attitudes, having groups separated by gender is much less relevant than separation by race.

In a recent study on families and health, one of the authors conducted two focus groups prior to developing the questionnaire. One group consisted of single parents only, and the other group included married people with children. The purpose of conducting the focus groups was to determine how different family structures, relationships, and experiences affect health care and decision making, and how family interactions and relationships influence health. Groups were asked, for example: How do you respond to your child's symptoms? Who takes the child to the doctor? Do you ask other family members or friends for advice? Does your job affect your ability to meet family needs? Is health an important issue in your family? What things do you do to promote good health? Of particular interest were the differences among participants and the words and phrases they used to describe their health and health behaviors.

When we either feel comfortable with the draft of the questionnaire or have tried to anticipate how respondents will answer some questions but are still unsure, it is time to try the questionnaire on real respondents. A pretest usually involves 20 to 40 interviews, the exact number determined by such things as the number of subgroups of interest, how these subgroups might interpret the questions, uncertainties we have about how well we are measuring our concepts, the time schedule, and the budget. After we complete a formal pretest and debrief the interviewers, we undoubtedly will need to revise the questionnaire and survey procedures. If the changes are extensive, we should conduct informal testing and then another formal pretest. The size of the second pretest is determined by the number of uncertainties we need to resolve. We keep repeating this process until we are completely satisfied with the questionnaire and the survey procedures.

The pretesting phase can last from 1 month to a considerably longer period. It is not uncommon for the pretesting phase, including the development of the final questionnaire and training materials, to last 2 to 3 months.

Stage 3: Final Survey Design and Planning

The pretest results should be used to improve the survey design and implementation plans. For example, in the pretest of a telephone survey of four physician specialty groups, one of the authors obtained response rates in the range of 40% to 50%. In subsequent focus groups with physicians, the discussion concentrated on what could be done to increase the response rates. One suggestion that was followed was to give the physicians a choice of how to respond: they could either complete a mail questionnaire or be interviewed by telephone. The rationale was to let each physician choose the method that was best for the physician's schedule. The suggestion was a good one because the overall response rate for the main study increased to 67%.

Pretesting may also help to decide how much time to allot between followup contacts in a mail survey or whether the final contact in a mail survey should be by mail or by telephone. There are very few hard and fast rules in survey research. Researchers need to be attuned to the factors that can affect response rates and data quality and be prepared to make adjustments.

During this stage, final changes should be made in the sampling plan, the questionnaire, interviewer-training procedures and materials, data-coding plans, and plans for analyzing the data. For example, in sampling we may learn that we need to select more phone numbers to get the desired number of completed interviews because we found more nonworking phone numbers than expected in the pretests. For the questionnaire, we may find that changing the order of some questions improves the flow of the interview, or we may be able to develop closed-ended response choices for some questions that were initially asked as open-ended questions. Another common occurrence is that we find the answers of a particular subgroup in the population, such as those older than age 55 years or black respondents, to be different from the responses of other subgroups or different from the responses expected; or we may find the number of completed interviews with these subgroups to be less than anticipated. Then we need to decide whether the sample sizes for these subgroups will be adequate or whether we need to oversample certain groups.

Stage 4: Data Collection

During this stage, we need to monitor the results of the sampling and data collection activities and begin coding and data file preparation. In sampling, we want to ensure that all the cases are accounted for and that those that have not yielded an interview are being worked thoroughly. Nonrespondents in telephone and face-to-face surveys should be recontacted on different days of the week and at different times of day. It is also important to monitor the results or the dispositions of the completed cases. What are the rates for refusals, noncontacts, ineligibles, and completed interviews? Are there any signs of data falsification? We typically validate between 10% and 20% of each interviewer's work by having a supervisor recontact respondents and/or edit completed interviews (cf. Johnson, Parker, & Clements, 2001). In the recontacts, a supervisor determines that the interview actually occurred, and also asks how long it took, because another way to falsify results is to omit asking some of the survey questions. The recontact is usually explained to respondents as a quality control procedure. In telephone surveys, extensive monitoring of interviews in progress typically replaces or supplements recontacting respondents. We also use biweekly or weekly sample-disposition reports to help spot or anticipate minor problems before they develop into major ones.

Monitoring the interviewers is equally important. Each interviewer's first few completed interviews should be thoroughly checked. We want to ensure that no questions are being missed, that complete information is being obtained, and that all instructions are being followed. It is important to provide feedback to the interviewers so that problems and mistakes can be corrected before they become established patterns. We should also monitor each interviewer's success at converting assignments to interviews. New interviewers are likely to accept respondents' comments of "I am too busy" or "I'm not interested in the topic" as refusals and to give up on the case. It is useful to discuss these situations in training sessions and to instruct interviewers on how to deal with them. If an interviewer has a disproportionate number of refusals, it is a good idea to examine each case with the interviewer and suggest possible ways of recontacting the respondents or ways to handle similar situations in the future. Good researchers attempt to minimize noncooperation by using specially trained interviewers to recontact refusals and convert them to interviews. These interviewers try to convince respondents who are not adamant refusers to reconsider and to grant them an interview. Many times these interviewers convert 20% to 40% of the initial refusals. Groves and Couper (1998) present a number of techniques that experienced interviews use to minimize refusals and noncontacts.

As data are being collected, we must code and enter the information from completed interviews into a computer data file. These activities are referred to as **data reduction** and are usually performed by a separate group of people. Prior to entering data into a data file, data-reduction staff edit the questionnaires. They look for the same information that the interviewer's supervisor was looking for: skipped questions, incomplete information, and incorrect, inconsistent, or illogical entries. This edit should occur as soon as possible after the interview, ideally within a few days and not more than 2 weeks after the interview. Time is important because the interviewer, respondent, or both may need to be contacted to resolve the problem. The sooner this contact occurs, the easier it is to recall the situation and correct the problem.

Stage 5: Data Coding, Data-File Construction, Analysis, and Final Report

The final stage of a survey includes coding and analyzing the data and writing a final report or papers describing the survey results. **Coding** is the assignment of numbers to the responses given to survey questions.[7] Coding respondents' answers to each question allows us to estimate characteristics or to look for patterns among variables. The following example illustrates coding.

In a telephone survey questionnaire used in a study of Maryland adults' attitudes about violent street crime, 824 interviews were completed (see Appendix B). Each respondent's answers were individually coded and entered into a data record, usually in the same order that the questions were asked in the survey. The procedures are the same whether the interviewing is done by computer-assisted telephone interviewing or by paper and pencil. A data record includes all the coded responses for one respondent. In addition, each respondent was given a unique identification number between 001 and 824, which was entered at the beginning of the respondent's data record.

When coding, every question or variable is given the same designated amount of column space within a person's data record. For each question or variable, every answer category is given a designated code number. In the Maryland Crime Survey, the first question asks about the crime problem in the respondent's neighborhood. There were four answer categories, coded 1 to 4, plus "don't know" which was coded as 8. If the respondent with ID number 001 answered question 1 "not very serious," the answer was coded as 3. If the answers to questions 2 and 3 were "very serious" and "stayed about the same," the code values would be 1 and 3, respectively. If the data record begins with the respondent's ID number, the code values for the first six columns of the first respondent's data record would be 001313. In the same manner, code numbers would be assigned to the rest of the first respondent's answers to the remaining 49 questions and to the answers given by the other 823 respondents. One simple rule should be remembered when writing and developing answer categories for questions: categories should always be mutually exclusive and exhaustive.

Before data analysis begins, the data are checked or "cleaned" to identify and correct coding and data-entry errors.[8] In the cleaning process, the coded response to each question or variable is checked for illegal code values and, when possible, for consistency with responses to other, related questions. If codes for respondents' sex, are 1 for males and 2 for females, then the cleaning process should disclose no code values of 3 or higher for that variable. Interitem consistency is checked in a similar manner. If a series of questions were asked only of male respondents, then we should verify that legitimate codes for these questions are entered for all the male respondents and that a code denoting "not applicable" is assigned to those questions for each female respondent. To complete all the data-reduction work, including coding, cleaning, and preparing a data file, typically takes a few days to 4 weeks after the last interviews are conducted, depending on the method of data collection and the survey's complexity. When the data are collected using a computer, these activities are moved "up front" and become part of questionnaire development.

The amount of time needed for data analysis and report writing depends on the survey's goals and commitments. After a clean data file is ready, it may take 4 to 6 weeks to write a report documenting the survey procedures and answering one or two research questions. If the plans are more ambitious, then more time must be allocated.

Example of a Time Schedule for a Study

The preceding overview of the stages of a survey gives you an idea of the number of decisions that must be made, the multiple factors that must be considered when making these decisions, the different activities that need to be done simultaneously, and, briefly, the amount of time some survey tasks require. In the example to follow, a more specific time schedule is illustrated. The study is a national RDD survey of 1,000 households with respondents who are 18 years of age or older. The topic of interest is biotechnology and food safety. (Exhibit 2.2 lists the major tasks in the survey and time allocated for each.) In the following discussion, we address key activities as if the survey is being conducted by computer-assisted telephone interviewing (CATI).

After we decide that a survey is appropriate, that it should be national in scope, and that a telephone survey using RDD is appropriate, we need to make an outline of the research questions and the kinds of survey questions we might ask respondents. While we are doing this, we need to search the literature for articles and books that report on related research. In reviewing

Exhibit 2.2   Time Schedule for a Study Involving 1,000 Random-Digit-Dialed
Interviews

Activity	Number of Weeks	Week Number
Review literature and draft questionnaire	8	1–8
Assemble think alouds with ten respondents	1–2	8–9
Revise questionnaire	2	10–11
Conduct pretest 1 (n = 25–40)	1	12
Debrief interviewers and revise questionnaire	3	13–15
Pretest 2 (n = 20–30)	1	16
Debrief interviewers, revise questionnaire, and develop training materials	4	17–20
Select samples (for pretests and main study)	12	8–19
Conduct main data collection	8	21–28
Code data and prepare data files	12	21–32
Analyze data and write report	Open	Open

these studies, we pay close attention to how the researchers defined and measured their concepts and to the specific questions used in their questionnaires. Specific attention should be given to what worked and what didn't. This information will be used in developing our questionnaire. Although we allocated 3 weeks for the initial literature search and review of past research, the perusal of literature for new studies or for ideas to solve problems can be an ongoing task. Drafting questions and organizing them into a questionnaire follows from this effort. We might spend 4 to 5 weeks writing questions and developing successive drafts of a questionnaire. We try each draft on a few colleagues or friends, make revisions, retry it on a few different people, revise it, and then try it again. While this is going on, those who will be doing the sample selection need to develop a sampling frame or purchase one from an academic or commercial organization.

Because not much social science research has been done on biotechnology, we do not know a lot about the public's knowledge of and attitudes about the topic. Thus, it is a good idea to assemble a few focus groups or think-aloud interviews to help in developing the questionnaire. The time schedule allocates 1 to 2 weeks for this task. Approximately 2 weeks prior to the focus groups or think-aloud interviews, respondents must be recruited.

In weeks 10 and 11 we will again revise the questionnaire and finish preparations for a formal pretest scheduled for week 12. The pretest will be done with a national, general-population sample. Thus, before week 12 we must select the sample, recruit and hire interviewers, and write and reproduce interviewer-training materials. Immediately after completing the pretest interviews, we will have a debriefing with the interviewers. This session may last half a day. We want to identify problems and discuss solutions, if possible.

If the interviewing is being done with CATI, key factors in scheduling are the nature of the CATI software and the experience of the person assigned to put the questionnaire into the system. Some CATI systems such as CASES (developed at the University of California, Berkeley) use a programming language that requires someone with programming experience to use. Other systems (e.g., Sawtooth) are less powerful in terms of the complexity of questionnaires they can handle, but are easier to use. It is important that there be a good match between the system and the programmer.

Closely related to the initial programming are debugging and reprogramming. Once the instrument is in the system, it needs to be "stress tested." That is someone needs to go through every path the interview can take to be sure the system performs as intended. Often errors will be found; time needs to be in the schedule to allow for correction (cf., Kinsey & Jewell, 1998).

Because we anticipate making major revisions to the questionnaire, we plan a second formal pretest of 20 to 30 respondents. The revisions and preparations for the second pretest will occur in weeks 13 to 15, with the second pretest conducted in week 16. Another interviewer debriefing will be held at the end of this pretest.

This pretest and debriefing may uncover no major problems or surprises, or it may reveal that revisions should be made to the questionnaire, requiring additional pretesting. Because a complex research problem or design may require further pretesting, this should not be a source of surprise or alarm. For our example, however, let's assume things went well. Over the next 3 or 4 weeks (weeks 17 to 20) final revisions will be made in the questionnaire and in the interviewer-training materials. At the same time, we will prepare for the selection and training of coders, develop coding materials, and write specifications for data cleaning and an analysis file. In weeks 8 to 19 we will conduct sample selection activities for the pretests and the main study. Before week 21, the questionnaires and related materials must be typed and reproduced. Because we plan on completing 1,000 interviews, we will need 1,400 or 1,500 copies of the questionnaire. If the questionnaire is 20 pages, 1,450 copies will amount to 29,000 sheets (58 reams) of paper for the questionnaire alone. We need to budget an adequate amount of time to type the final copy of the questionnaire; reproduce, collate, and staple the copies; and then integrate the training materials with the questionnaire and sample assignments along with other administrative materials.

If using CATI, the activities of weeks 17 to 20 are still appropriate. Additionally, after pretesting and questionnaire revision, the revised instrument needs to be CATI tested. This is an entirely separate testing protocol from testing the content of the questionnaire. It requires additional time and special skills. If the testing is short-changed, the risk is that the instrument will fail during data collection, a very expensive breakdown. If you have not worked with a system before or do not have an experienced programmer available, then learning the software is a totally separate task from designing and implementing the survey. The same cautions apply to computer-assisted personal interviewing (CAPI) and other computer-assisted interviewing (CAI) applications: they are powerful, but require a particular expertise to use efficiently.

In week 21 we will train the interviewers and begin data collection. We have allocated 8 weeks for data collection. A number of factors affect the time allocated to data collection. The time period must be long enough so that we will have a reasonable chance to interview each sample case. We want to allow enough time to reach people who are away on vacation, have very busy schedules, are away from home a lot, are ill or in a short hospital stay, and so forth. Each case must be thoroughly worked so that the sample will represent the diverse nature of the population. We must also consider the interviewers. How many hours a week can we ask them to work? How many interviews can we expect them to complete in an average week? If they work in a central telephone center, are there enough cases to keep them busy for a full shift? If we hire and train 15 interviewers, ask them to work an average of 15 hours per week, and if none leave during the data collection period, then each interviewer must complete about 8.3 interviews per week. This goal is easy to achieve and, if no major difficulties occur, data collection will be completed in less than 8 weeks.

The same week interviewers are trained, we begin training the coders and writing the specifications for the analysis file. Coding should be concurrent with interviewing; thus, it is important for coders to edit questionnaires as soon as possible. Our schedule allocates four weeks after data collection has ended to complete coding, data cleaning, and preparing the data analysis file. If there are only one or two open-ended questions and the interviewers do a thorough job, it may take less than 4 weeks to produce a complete data analysis file.

The time required to complete a survey depends on a number of factors, including project complexity, sample size, method of data collection, amount and complexity of the information collected, analyses requirements, as well as other considerations. For example, response rate can be affected by the length of the data collection period. More callbacks and refusal conversion can be done; tactics such as mailing refusal conversion letters (especially if incentives are included) can be effective. All these procedures take time. Other things being equal, a longer time period increases the chances to build up the response rate. Conducting a high-quality survey from start to finish requires hundreds of decisions and a number of stages, and it takes time. In our example, the conceptualization of a research idea, the project implementation, and the creation of a data analysis file take approximately 8 months and reflect the amount of time and level of effort required to properly conduct a moderate-size survey without cutting corners. (This 8-month schedule does not include any data analysis or report writing.) However, we don't always have the luxury of an adequate amount of time to do a survey. Some circumstances require that a survey be conducted more quickly: the need to determine public reaction to a news event (e.g., a military invasion or presidential assassination attempt), or the requirement to complete a survey as part of a class project.

When time is short, the researcher must (a) recognize that by cutting corners the data may have shortcomings that affect the reliability and validity of the results, and (b) determine which compromises may have the least effect on the results. To minimize compromised results, a researcher may, for example, want to buy a sample from a commercial organization rather than to develop and select the sample herself. Or, the researcher may want to rely on questionnaire items used in previous surveys and forego a number of pretesting procedures for assessing the validity of the items in their survey population. In both situations, the researcher assumes that things will work well and must be willing to accept the risks involved.

Notes

1. We will learn in Chapter 8 that there is a simpler and better sampling frame for this type of telephone survey.

2. One reason for these differences may be question wording. The Census asked, "Is there telephone service available in this house, apartment, or mobile home from which you can both make and receive calls?" The CPS asks a series of six questions because the goal is to conduct interviews 2 to 4 and 6 to 8 by telephone. The basic question is, "Since households included in this survey are interviewed (again/again during the next 3 months), we attempt to conduct the followup interviews by telephone. Is there a telephone in this house/apartment?"

3. In general, cellular exchanges are distinct from landline exchanges. Most RDD surveys do not include cell phones. The reasons are (a) cost—the cell phone owner and caller both pay for the call. Most cell phone owners are not interested in paying to be interviewed, thus refusal rates are extremely high. (b) Liability issues—there is concern that if a respondent is driving her car while being interviewed and is involved in an accident, the survey organization may be a contributing factor. (c) Selection probabilities—cell phones comprise samples of individuals while landline

households are samples typically of multiple individuals. The data need to be weighted when combined. We discuss this in later chapters.

4. Households in transition include people who have moved recently; immigrants; the unemployed; people who are divorced, separated, or widowed; and so forth.

5. Most survey researchers do not encourage or support the copyrighting of survey questions. The standard procedure is to freely encourage the use of the same questions in order to test the reliability and validity of the items with different populations and at different times.

6. In Chapter 6 we discuss other procedures for testing questions and assessing the results, such as expert panels, postinterview interviews, taped interviews with behavior coding, and respondent debriefing.

7. Sometimes letters or symbols are used to code responses. This is the exception rather than the rule, and it does not allow statistical analysis of the data. We ignore this type of coding.

8. When data are directly entered into a computer, for example, through the use of computer-assisted telephone or personal interviewing, coding and cleaning instructions are written after the questionnaire is finalized and before main collection begins. Thus, cleaning and coding checks are done as part of the interview.

3

Selecting the Method of Data Collection

O nce we have decided that a survey is the appropriate method for investigating the research problem, the next question to consider is what data collection method is most advantageous. Until recently, the three most common survey approaches were mail, telephone, and face-to-face. In the last decade, there has been a proliferation of computer-assisted survey methods and rapid development of a fourth approach, surveys conducted via the World Wide Web, or Internet surveys.[1] Combinations of any of these are also possible, and we discuss them later in the chapter. A fifth approach is a group-administered questionnaire, such as one completed by all students present in selected classrooms or classes on a particular day. This type of survey is usually conducted in conjunction with one of the other survey methods and is also discussed later, under combinations of methods. We begin with a brief description of the four main survey approaches.

Evaluating the Advantages and Disadvantages of the Four Survey Methods

There is no one "best" survey method; each has strengths and weaknesses. It is important to know what these are and to evaluate research objectives with reference to the advantages and disadvantages of each method. The decisions involved in selecting a method of data collection must be made study by study.

In choosing a method of data collection, three broad categories of factors, with many subcategories, must be considered:

1. Administrative or resource factors

2. Questionnaire issues

3. Data-quality issues

Exhibit 3.1 summarizes the comparative strengths and weaknesses of mail, Internet, telephone, and face-to-face surveys. In deciding on a method, we must consider questions relevant to each category of factors. Regarding administrative and resource factors, we need to consider how much time there is to do the research and how much money we have for hiring interviewers and/or coders; purchasing hardware, software and supplies; whether we should use incentives; and buying or constructing a list of the population that we want to sample and interview. Questionnaire issues include how many and what kinds of questions need to be asked to adequately measure the concepts and achieve the research objectives. We must decide if one method of collecting the data is more cost-effective or yields fewer reporting errors than other methods. In considering data-quality issues, we must ask whether more respondents are likely to cooperate with one method of data collection than with another, whether we get more accurate or more complete answers if we use interviewers, and whether one method is more likely to include the population we want to study.

Many researchers start by evaluating the feasibility of a mail, Internet, or telephone survey. There may, however, be major, compelling reasons for doing a face-to-face survey; for example, respondents need to be shown lists of answer categories for key questions, the target population has low telephone or computer coverage, or the respondents are likely to have difficulty reading or writing answers to questions. The first two questions we need to ask are "Who are the survey respondents?" and "Is my research question more amenable to one method than to another?" If the answer to the second question is no, then time and money are the next considerations. In addition to thinking about time and money, researchers need to assess several other important factors, as Exhibit 3.1 indicates. The following sections discuss the advantages and disadvantages of each method with specific reference to the factors listed in Exhibit 3.1.

Exhibit 3.1 Comparison of Major Survey Methods

Aspect of survey	Mailed Questionnaires	Internet Surveys	Telephone Interviews	Face-to-Face (in home) Interviews
Administrative, Resource Factors				
Cost	Low*	**Very Low**	Low/medium	High
Length of data collection period	Long (10 weeks)	**Very Short/short** (1–3 weeks)	**Short** (2–4 weeks)	Medium/long (4–12 weeks)
Geographic distribution of sample	**May be wide**	**May be wide**	**May be wide**	Must be clustered
Questionnaire Issues				
Length of Questionnaire	Short/medium (4–12 pages)	Short (<15 minutes)	Medium/long (15–35 minutes)	**Long** (30–60 minutes)
Complexity of questionnaire	Must be simple	**May be complex**	**May be complex**	**May be complex**
Complexity of questions	Simple/moderate	Simple/moderate	Must be short and simple	**May be complex**
Control of question order	Poor	Poor/fair	**Very good**	**Very good**
Use of open-ended questions	Poor	Fair/good	Fair	**Good**
Use of visual aids	Good	**Very good**	Usually not possible	**Very good**
Use of household/personal records	**Very good**	**Very good**	Fair	Good
Rapport	Fair	Poor/fair	Good	**Very good**
Sensitive topics	**Good**	Poor/fair	Fair/good	Fair(Good with A-CASI)
Nonthreatening questions	Good	Good	Good	Good
Data-Quality Issues				
Sampling frame bias	Usually low	Low/high	Low (with RDD)	Low
Response rate	Poor/good	Poor/good	Fair/good	**Good/very good**
Response bias	Medium/high (favors more educated people)	Medium/high (favors more educated people)	**Low**	**Low**
Knowledge about refusals and noncontacts	Fair	Fair	Poor	Fair
Control of response situation	Poor	Poor	Fair	**Good**
Quality of recorded response	Fair/good	Fair/good	**Very good**	**Very good**

*Boldface indicates that the method has an advantage over one or all of the other methods in the specific survey component noted.

Mail Surveys

Mail surveys involve sending a brief prenotice letter and then a detailed cover letter and questionnaire to a specific person or address (Dillman, 2000). The detailed cover letter should state the purpose of the survey, who is the sponsor or is collecting the data, who is to complete the questionnaire, why a response is important, an assurance of confidentiality, and when to return the questionnaire (usually within 10 days). Mail surveys require a questionnaire that is totally self-explanatory; the importance of clear and simple statements cannot be overstated, because instructions and questions must be uniformly understood by a wide variety of respondents. If respondents do not understand the task or view it as difficult or too time-consuming, they will probably not complete the questionnaire or will make errors. Cover letters provide a telephone number for respondents to call if they have questions about the legitimacy of the survey or have difficulty interpreting any of the questions. Our experience, however, indicates that very few respondents call, probably fewer than 1%, even if the survey contains controversial or personal items. When respondents have doubts or concerns, they are more likely to not complete the questionnaire or to skip unclear items than to call with questions. Well-designed surveys strive for high response rates by sending a thank-you postcard, a new cover letter and another copy of the questionnaire, and a final "special contact" to nonrespondents by telephone, Priority Mail or FedEx (Dillman, 2000). The use of monetary and nonmonetary incentives in the initial mailing has been found to be effective (Church, 1993).

Advantages

The mail survey is significantly less expensive than a telephone or face-to-face survey. It requires money for postage and envelopes; a list of the names and addresses of the study population (a sampling frame); money to type and print a professional-looking questionnaire; and people to assemble the materials to mail out, keep track of who responds and who does not, do remailings, edit and code the returned questionnaires, enter the data into a computer file, and construct an analysis file.

A major difference between mail and telephone or face-to-face survey methods is that, in a mail survey, the data are not collected by interviewers. This characteristic has both advantages and disadvantages. A clear advantage is cost. The cost of first-class postage to mail out and return an eight-page questionnaire (printed on four sheets) with a return envelope is minimal; an interviewer's pay rate can be $8.00 per hour and higher. Another advantage is that because the cost of mailing a questionnaire is the same whether we send it to someone in Alaska or across town, we can cost-effectively survey a national sample of respondents, which is more diverse than a citywide or statewide sample.

The questionnaires used in mail surveys also have several advantages. One advantage is that respondents may consult household or personal records. For example, if we need to know the dollar amount of interest that respondents pay on a home mortgage or how much they paid in doctors' fees within the last 6 months, we can encourage them to consult records rather than to answer from memory. The result is greater response accuracy. A second advantage is that we may use visual aids. If we want respondents to see a list of possible answers before responding, we list the answers with the question. If we want respondents to define the geographic boundaries of their neighborhood or to indicate their travel route to work, we can reproduce a map on the questionnaire or on a separate enclosed sheet.

Mail surveys have also been successful in the collection of data about sensitive topics. Respondents seem to be more comfortable answering questions about intimate medical problems (Hochstim, 1967), personal bankruptcy, convictions for drunk driving (Locander, Sudman, & Bradburn, 1976), and other potentially embarrassing topics in self-administered questionnaires (Aday, 1996; Tourangeau, Rips, & Rasinski, 2000). For these types of topics, the more anonymous the method of data collection, the higher the rate of reported behavior. Keep in mind, however, that while a mail survey is better than other methods, it is far from perfect. Many sensitive behaviors remain unreported, even in anonymous mail surveys (Sudman & Bradburn, 1982).

The length of time to conduct a mail survey is fairly constant—usually 8 to 10 weeks—regardless of the sample size and its geographic distribution. As mentioned earlier, we must allow enough time for the questionnaires to get to the respondents, for respondents to complete them, for the return of the questionnaires, and for followup mailings and returns. The time required may be either an advantage or a disadvantage, depending on the particular survey. If we have a sample of 2,000 or more, a mail survey will probably take the same or less time than a telephone survey. On the other hand, a telephone survey of 300 to 500 respondents typically can be completed in 2 to 4 weeks, as opposed to the 8 to 10 weeks that a mail survey requires.

Before we discuss why response rates to mail surveys can vary widely, it is necessary to define the term. We define **response rate** as the number of eligible sample members who complete a questionnaire divided by the total number of eligible sample members.[2] The word *eligible* is crucial to the definition because occasionally people we do not want to include in our survey may receive a questionnaire and may respond. For example, assume we want to measure the attitudes of college males toward abortion. If one of our questionnaires was sent to a female and she responded, we would treat that response as ineligible. In computing our response rate, we would not count the interview and we would reduce the sample size by one. Response rates are an important measure of survey quality; typically the higher the rate, the better the quality. Of course, a 30% response rate is not much better than a 20% response rate; both are unacceptably low.

We can achieve reasonably high response rates to a mail survey when the topic is highly salient to the respondent. For example, a survey sent to physicians asking about the burdens of malpractice insurance and actions the government can take regarding the situation would likely elicit a good response. The same survey would not do as well with the general population. Why? Because, although many people in the general population may think the issue is important, they are not nearly as concerned about it as physicians are.

Two effective means of increasing mail response rates are incentives and repeated follow-up attempts (Yammarino, Skinner, & Childers, 1991; James & Bolstein, 1990). Various types of incentives have been used, including different amounts of cash or checks, books, pens, key rings, tie clips, lottery and raffle tickets, contributions to charities, and other assorted materials or promises. Church (1993) reports that prepaid monetary and nonmonetary incentives do increase response rates, with monetary incentives having the greatest effects, 19.1% and 7.9%, respectively. Incentives that are conditional on first returning the completed questionnaire did not have an impact. The key is to include the incentive with the initial request.[3]

Disadvantages

Response bias occurs when one subgroup is more or less likely to cooperate than another. For example, mail surveys are subject to response bias because they do not achieve good response rates from people with low education, people who do not like to write, those who have difficulty reading, and those who do not have an interest in the topic. Response bias is potentially greater in mail and Internet surveys than with other methods because respondents can more easily ignore these questionnaires than a polite but persistent interviewer. For this reason, the cover letter and the appearance of the questionnaire are critically important in encouraging cooperation and minimizing response bias. Usually, our only contact with the respondents is through these materials, and, therefore, they must be especially convincing and appealing.

To assess potential response bias, we need to know as much as possible about the nonrespondents (refusals and noncontacts). It is important to determine whether those who refuse or who cannot be reached for an interview (noncontacts) are different, in terms of our dependent variable(s), from those who cooperate.[4] When we have a list of our population, we have information that can be used to determine potential response bias. For example, from names and addresses, we can determine if males or females were more likely to cooperate and if those living in large cities were more or less cooperative than people in other areas. Our list may also contain other information to help identify response bias.

A number of characteristics of mail surveys make them less effective than surveys administered by interviewers. Because respondents can look over the questionnaire before deciding whether or not to complete it, the questionnaire cannot be very long, nor can it look complex or difficult to complete. In a mail survey, nonresponse is more strongly related to interest in the topic than in telephone surveys where the decision to participate or not is made before knowing very much about the survey content.

A mail questionnaire must be completely self-explanatory because no one is present to assist if something is confusing or complex. Even experienced researchers often think that a statement or question is self-explanatory when, in fact, a significant segment of the study population misinterprets it. A lack of understanding by respondents can and does affect the quality of survey responses. It is easy for respondents to skip questions they do not understand or do not want to answer. In addition, the researcher has little control over the order in which respondents answer the questions or over who actually fills out the questionnaire. Finally, the answers to open-ended questions are less thorough and detailed in mail surveys than they are in surveys administered by interviewers. Unless prompted by interviewers, many respondents provide only the minimum amount of information needed to answer the question. This tendency is especially true of respondents with low levels of education and of those who do not like to write. Very few field experiments have been conducted on responses to open-end questions between self-and interviewer-administered questionnaires. Two testable hypotheses are feasible. For self-administered questionnaires, one might expect longer and more complete answers because of fewer time constraints, whereas one might expect the same result from interviewer-administered surveys because of the opportunity to probe the respondent's initial answer (Groves et al., 2004). A study by de Leeuw (1992) found no differences between self and interviewer-administered surveys in length of response to four open-ended questions. Future studies need to disentangle method of data collection from respondent's interest in the topic controlling for socioeconomic status.

Internet Surveys

Internet surveys are a relatively new and increasingly popular form of self-administered survey with many similarities to mail surveys, but also some important differences. Because of the coverage issues discussed below—only about half of the U.S. population currently has Internet access (Federal Communication Commission [FCC], 2003)—probability surveys via the Internet typically involve list samples of known or likely Internet users. Depending on what information is available from the list sampling frame, potential respondents may be contacted by telephone, regular mail, or e-mail. The purpose of the initial contact is to explain the purpose and importance of the survey, identify the sponsor, provide an assurance of confidentiality, and provide instructions for accessing the survey Web site. To insure that only sampled individuals complete the survey and that they complete it only once, each person is given a unique personal identification number (PIN) that must be entered to access the online questionnaire. The PIN is either included in the initial contact letter or it can be embedded in an extension to the URL in an e-mail. In the latter situation, the respondent clicks on the URL, the Web site reads the PIN, avoiding the need (and errors) for respondent entry.

A similar, but brief introductory message stating the purpose of the survey and encouraging participation should be provided on the first screen or welcome page of the Web questionnaire along with straightforward instructions for entering the PIN, if appropriate, and getting to the first page of the questionnaire. The welcome page should also include an e-mail address and/or telephone number for respondents who wish to ask questions about the survey or who have difficulty responding. It is also a good idea to include a mailing address for respondents who prefer to print out a copy of the questionnaire and return it by mail. The online questionnaire must be carefully designed to be accessible and understandable to respondents with different levels of education, computer literacy, computer hardware and software, and Internet access. Multiple reminder contacts to sampled individuals who do not log on to the survey Web site or submit completed questionnaires within specified periods of time, delivered via e-mail, if possible, are essential for increasing response rates.

Advantages

The two great advantages of Internet surveys are the low cost and the speed of data collection. Internet surveys eliminate not only the interviewer costs of face-to-face and telephone surveys, but also the paper, questionnaire reproduction, postage, and data entry costs of mail surveys. Moreover, the geographic distribution of the sample and sample size has almost no effect on the cost of an Internet survey. Collecting data from a national or even an international sample via the Internet is no more costly than surveying a geographically concentrated sample. Because data collection costs constitute such a high proportion of the total cost of a survey, Internet surveys, with their extremely low data collection costs, allow researchers to increase sample sizes for relatively small overall cost increases. More importantly, for a given sample size, much more follow-up can be done for little additional cost.

Speed of data collection is the second great advantage of Internet surveys over other survey methods, especially mail and face-to-face surveys. The data collection period for Internet surveys is typically 10 to 20 days and, in some cases, may be significantly shorter. For example, in an Internet survey of University of Michigan students, 30% of the total number of completed questionnaires were received by the end of the first day and 50% of all completes had been received by the third day (Couper, Traugott, & Lamias, 2001). Maximizing response rates, however, requires that time be allowed for sending reminder messages and for receiving completed questionnaires from less-speedy respondents.

Self-administered online questionnaires have several advantages over questionnaires for mail, telephone, or face-to-face surveys. Web questionnaires may include complex skip patterns because the skips are programmed into the questionnaire and implemented automatically—they are transparent to the respondent; a variety of visual aids such as pop-up instructions, drop-down lists, pictures, video clips, and animation, even audio. In an online questionnaire, it is also possible to incorporate the response to an earlier question into a subsequent question. The caveat associated with all these innovative possibilities is that the researcher must be careful to minimize potential response bias by designing for the "lowest common denominator," that is, for respondents who do not have up-to-date computer equipment and who have slower telecommunications access. If a questionnaire is so complex that it takes an inordinate amount of time to download, respondents will either give up completely or submit partially completed questionnaires.

The ability to obtain reasonably complete and detailed answers to open-ended questions may prove to be an advantage of Internet surveys over mail surveys. There is limited evidence that e-mail survey respondents provide more complete answers to open-ended questions than do mail respondents (Schaefer & Dillman, 1998). However, it is unclear whether this is a result of the respondent's level of education or of the method of data collection.

An interesting thesis from research on human–computer interactions, called *social interface theory*, postulates that "humanizing cues in a computer interface can engender responses from users similar to human–human interactions" (Tourangeau, Couper, & Stieger, 2001). If true, humanizing cues could yield both advantages and disadvantages in Internet surveys. The advantage would be that virtual interviewers (e.g., talking heads) or other personalizing techniques could be used to establish a greater sense of rapport with the resulting benefits associated with interviewer-administered surveys. On the other hand, there is concern that humanizing cues may produce social desirability effects in surveys dealing with sensitive or potentially embarrassing issues such as sexual practices, alcohol and drug use, voting behavior, and church attendance. However, the good news is that a recent experimental study designed to explore the effects of human interface features and personalization techniques on Internet survey responses found that "neither the level of personalization nor the level of interaction had much effect on reports about sensitive topics" (Tourangeau et al., 2001, p. 7).

Disadvantages

By far, the biggest disadvantage of Internet surveys is that a large proportion of the U.S. adult population does not have Internet access. The September 2001 Current Population Survey estimates that 56.5% of U.S. households have computers and 50.5% have access to the Internet (FCC, 2003). In spite of the rapid growth in Internet use, it will be many years before coverage approaches the level of telephone coverage in the United States and some experts question whether it ever will. As a result, it is not possible to conduct researcher designed and implemented general population probability surveys via the Internet; however, commercial firms are reporting this possibility (Huggins & Eyerman, 2001). The lack of good sampling frames, even for those who have Internet access, is a disadvantage. Researcher-conducted, probability-based Internet surveys are possible for only a very limited number of special populations for which reasonably complete and accurate sampling frames are available, such as university students and faculty, federal government employees, employees of certain companies or corporations, and members of some professional organizations.

Low response rates and the resulting potential for response bias are also significant disadvantages of Internet surveys. Although information from probability surveys conducted via the Internet is scarce, it appears that response rates are typically lower than for mail surveys. For example, an Internet survey of university students achieved a 41% response rate (47% if partial interviews are included), which the researchers characterize as similar to other similar Internet surveys (Couper et al., 2001). In a study of hospitality professors that was designed to compare mail, fax, and Web-based survey methods, Cobanoglu, Warde, and Moreo (2001) obtained a response rate of 44% in the Internet survey. Interestingly, the results of this study are contrary to the contention that Internet survey response rates are generally lower than response rates for mail surveys as the mail survey response rate was only 26%. Exceptions notwithstanding, we agree with the observation that ". . . much work remains to be done to bring Internet survey participation rates up to the levels of mail surveys of similar populations" (Couper, 2000). Undoubtedly, some of that work will need to focus on developing the kinds of motivating features that have been successful in increasing mail survey response rates, such as personalized cover and reminder letters and the use of incentives. Overcoming the technical difficulties that result in low response rates (e.g., slow telecommunications speeds, unreliable Internet connections, and low-end browsers) will be even more challenging.

Response bias is also a concern in Internet surveys. Again, research is limited, but it appears that people with lower education levels, those who have minimal experience using computers, and those with older computer equipment and low-end browsers are less likely to complete an online questionnaire than are highly educated, computer literate individuals with more up-to-date equipment. In addition to education, computer literacy and quality of computer equipment, one research study found that minority students had lower response rates than their white counterparts (Couper et al., 2001). The researchers point out that they were unable to determine whether the difference was a result of the topic of the survey (affirmative action) or of differing levels of computer use and familiarity between the two groups. Clearly, this is an area where future research will be important.

Internet surveys share several characteristics of mail surveys that make them less effective than interviewer-administered surveys. First, online questionnaires must be relatively short. To avoid high rates of nonresponse, item nonresponse, and breakoffs, researchers who have experimented with varying questionnaire lengths have found that online questionnaires should take no longer than 15 minutes to complete and a maximum length of 10 to 12 minutes is much preferable (Couper, 2001). Second, like a mail questionnaire, an online questionnaire must be completely self-explanatory because there is no interviewer to explain confusing or complex instructions or questions. Third, just as a mail questionnaire should appear the same to all respondents, so should an online questionnaire. Unlike a mail questionnaire, however, it is not always easy to design an online questionnaire that appears identical to all respondents. Questionnaire designers must understand the factors that affect the appearance of an online questionnaire—size and screen resolution of computer monitors, operating systems, browsers, transmission speeds, and the like—and minimize as much as possible the differences in question appearance caused by these factors. Fourth, although it is possible to design Internet surveys so that respondents must answer each question in the order presented, that practice is strongly discouraged. If respondents are required to answer each question before proceeding to the next, those who encounter questions that they choose not to answer or that they legitimately cannot answer will most likely exit the survey, thereby increasing the nonresponse rate. Forcing respondents to answer each question before proceeding or allowing them to check a "don't know" or "refused" category, provides them different options than are given in other types of surveys (Dillman, 2000). Thus, in a well-designed Internet survey, the researcher does not control the order in which respondents answer the questions. Finally, a researcher conducting either a mail or an Internet survey has no control over who actually completes the questionnaire or over the response situation (i.e., whether the respondent completes the questionnaire at home, at work, at an Internet café, or some other public place, or whether others are present while the respondent is answering the questions).

Some preliminary research suggests that Internet surveys, unlike other types of self-administered surveys, may not be appropriate for collecting data on sensitive topics. It appears that respondents' concerns about the security of the Web may outweigh the anonymity of the self-administered format (Couper, 2000; Dillman, 2000; Cho & LaRose, 1999). Such concerns are likely to result in high nonresponse rates to surveys on sensitive topics, high item nonresponse for sensitive questions, less-honest reporting, or some combination of all three. More research is necessary to determine what topics respondents consider sensitive and how they deal with specific questions. The following Web site keeps abreast of Web survey methodology: http://www.websm.org.

Telephone Surveys

In a telephone survey, telephone numbers are selected in a variety of ways. Numbers can be selected randomly from a phone book or created by using an existing telephone number and dropping one or more of the last four digits and substituting random numbers. Respondents themselves can be selected from a list, for example, a membership directory that includes telephone numbers. The interviewers are trained in procedures for contacting respondents and administering the questionnaire (Guensel, Berckmans, & Cannell, 1983). They are given an assignment of respondents or telephone numbers to call for interviews and are trained to collect information in a uniform and consistent manner. They are to ask questions exactly as written and in the same sequence for all respondents. The reliability of the data depends partly on following such standardized procedures. Since the mid 1990s, telephone surveys have faced a number of technological challenges which we discuss in the following pages.

Advantages

The telephone survey is the most widely used survey method today. This method has increased in use over the past few decades as a number of limiting factors or concerns have been overcome or shown to be overstated.

One factor contributing to the popularity of telephone surveys is the increase in the number of households with telephones. In 1960, 80% of households had a telephone; in 1970, 83%; in 1980, 92%; in 1990, 94.8% and in 2001, 95.5% (Frey, 1989; FCC, 2003). Thus, coverage error (households without telephones) is small.

Also of concern to researchers is the number of households with unlisted telephone numbers, which varies by community. When the proportion of these households is high, excluding them can be a serious omission. Survey Sampling, Inc. (Piekarski, 1997) estimates that some metropolitan areas have few households with unlisted telephone numbers; for example, 6.4% of households in Panama City, Florida, 7.8% in Sheboygan, Wisconsin, and 13.4% in Enid, Oklahoma. In other places, a high proportion of households have unlisted telephone numbers, for instance, 38.4% of households in San Antonio, Texas, and 64.4% in San Francisco, California. The metropolitan area with the highest proportion of unlisted telephone numbers was Stockton–Lodi, California, with 73.4% of households; in fact, 23 of the 24 highest metro areas with unlisted telephone numbers were in California. Anchorage, Alaska, with 67.6% of households unlisted, was the sole non-California city. RDD has allowed researchers to overcome this problem, but it costs appreciably more to do an RDD survey than to do a survey of respondents selected from a telephone directory.

Researchers have also been concerned about the quality of data collected from a telephone survey. A number of studies show that phone surveys compare favorably with other methods and that the limitations stem from the type and amount of information that must be collected. For example, if respondents must consult records, or if the interview will take much longer than 35 minutes, a telephone survey may not be the best method. However, if the interviewer asks mostly attitudinal and behavioral questions that can be answered accurately from memory, and the interview takes 30 minutes or less, then a telephone survey is a viable approach.

Telephone surveys are intermediate in cost between mail and face-to-face surveys. For a telephone survey, we need

- Interviewers
- A centralized work center with telephones and space for the interviewers to work
- Monitoring equipment, if possible
- A sample of the population, developed or purchased
- Staff to keep track of interview results, to edit and code completed interviews, and to build a computer data file for analysis, or to program the questionnaire, sample information, and write the data file and cleaning specifications in the CATI software.

Interviewers are essentially proxies for the researcher: We hire people to do interviews for us rather than doing the interviewing ourselves. This has both advantages and disadvantages. On the one hand, the research can be completed much more quickly than it could if the researchers were working alone. On the other hand, interviewers must be selected, hired, trained, and supervised. We must ensure that each interviewer is reading the questions verbatim, coding the responses accurately, and not biasing the survey in any way. Interviewers must also be paid. Both telephone and face-to-face methods are labor intensive, which is the major reason they are more expensive than mail and Internet surveys.

There are a number of clear advantages to telephone surveys. Because the Current Population Survey (CPS) estimates that 95.5% of households in the United States have telephones, **sampling frame bias** is low for RDD surveys. (However, rural areas, some areas in the South, and economically depressed areas are the areas most likely to have lower percentages of telephone households. There is also a growing proportion of households with cellular only telephones which are not included in many sampling frames. The 95.5% probably includes cell phone-only households.) Also, many people are willing to be interviewed by telephone, another reason for low response bias. However, some foreign-language groups are not comfortable being interviewed on the telephone, and survey success may necessitate that respondents and interviewers match by nationality and that interviewers speak the respondents' language.

Response rates for telephone surveys are usually in the range of 40% to 80% when repeated callbacks are used to reach people who are difficult to contact. We usually recommend 6 to 15 callbacks on different days of the week and at different times of the day. Response rates are usually better for telephone surveys than for mail and Internet surveys, primarily because an interviewer personally attempts to convince the respondent of the importance of the research, schedules the interview at a convenient time for the respondent, and reads and records the respondent's answers. Thus, the use of interviewers reduces the burden on the respondents considerably.

Location also affects response rates: In an inverse relationship to the size of the cities being surveyed, better response rates are achieved in small towns and cities than in the larger metropolitan areas.

Another advantage is that the length of the data collection period is usually as short or shorter for telephone surveys than for most other methods. Telephoning is a quick way to initiate contact with a person or household and to make callbacks, an important factor in contacting difficult-to-reach people.

The geographic distribution of the sample can be wide because it is easy and relatively cheap to purchase a sampling frame of all the area codes and telephone prefixes in the United States or a list-assisted sample. Nevertheless, the cost of a statewide or national survey is slightly higher than that of a local city survey because the cost of purchasing and/or selecting the sample is higher and we must pay long-distance charges. Costs for interviewing time are about the same for local, statewide and national surveys. The main difference is that the project needs to be staffed over more hours in a national survey to cover all time zones than is the case for local or state surveys.

The use of interviewers provides a number of advantages to the efficacy of the questionnaire and quality of the data. With thoroughly trained interviewers, the quality of recorded answers should be very high. In addition, because the interviewer is doing all the work—reading the questions and recording answers—interviews can last more than 30 minutes if the subject matter is interesting to the respondent. Also, because interviewers are trained in how to ask each question and in what order, the design or layout of the questionnaire can be complex. For example, the questionnaire can employ multiple skip patterns, in which responses to certain questions determine what other questions are asked or skipped by respondents. Well-trained interviewers can handle a variety of situations and formats; they can, for example, probe respondents' answers that are not clear.

Another advantage to using interviewers is that they are able to control the order of the questions. Respondents surveyed by telephone have no idea of what question will be asked next or of the number of questions in the survey. If they saw how many questions were in the survey or that a "yes" response would result in being asked a series of followup questions, they might be less inclined to be interviewed or they might answer "no" to the first question of the series.

Finally, it is possible for interviewers to establish rapport with respondents over the telephone and thus convince them to complete the interview, to believe in the authenticity and relevance of the research, and to provide complete and accurate answers to even sensitive questions. Telephone surveys are comparable to the other methods in eliciting responses to nonthreatening questions; they are also quite good for sensitive topics such as sexual behaviors that may lead to certain diseases (Czaja, 1987–1988), acquired immunodeficiency syndrome (AIDS) and high-risk behaviors (Binson, Murphy, & Keer, 1987; Catania et al., 1996), and other similar topics.

Disadvantages

Telephone surveys do have limitations. Technologies such as cell phones, caller ID, pagers, fax machines, Internet access, modems, answering machines, call blocking, and call forwarding have more than doubled the number of area codes and telephone numbers in the 1990s, making it more difficult and time-consuming to contact a household and its occupants. At the same time, the use of the telephone for telemarketing became so bothersome that the Federal Trade Commission now maintains a National Do Not Call Registry. Although telephone surveyors are exempt from registry restrictions along with charities, political organizations, and companies with which the respondent does business, the registry's creation is a clear indication that the general public is increasingly averse to unsolicited telephone calls. As a result of these developments, the average number of telephone contact attempts required to complete an interview has increased and survey response rates have dropped significantly.

Telephone survey questions must be short and fairly simple, and the answer choices that are read to the respondents must also be few, short, and simple. Otherwise, respondents may not keep all the information in mind. Sentences should be limited to 20 words or less, language kept simple, and no more than 4 or 5 short-answer categories read per question (Payne, 1951). When sentences are long or answer categories are numerous, respondents may remember only the first parts or the last parts. These are called *primacy* or *recency* effects (Schuman & Presser, 1981). A notable exception is that Sudman and Bradburn (1982) suggest using long, wordy questions when asking about sensitive behaviors.

The inability to use visual aids such as pictures, product samples, or lists of response alternatives for questions with many answer categories can also be a disadvantage. Various methods have been tried to get around this limitation with generally poor results. For example, if we have the respondents' names and addresses, we can mail materials to their homes beforehand. Similar to a combination mail and telephone survey, this method necessitates sending a cover letter explaining the materials before the telephone interview is carried out. This combined method is somewhat possible with RDD surveys by purchasing the names and addresses of the sample respondents that have listed telephone numbers. However, the use of advance letters with the listed portion of a RDD sample has yielded mixed results (Parsons, Owens, & Skogan, 2002; Singer, Van Hoewyk, & Maher, 2000).

A telephone-mail-telephone sequence was reported by Kanninen, Chapman, and Hanemann (1992), but it resulted in a very low final response rate. The survey asked respondents about their willingness to pay for five possible environmental protection programs in California's San Joaquin Valley. An RDD sample of phone numbers were called and eligible households were identified. During the initial telephone call, the purpose of the survey was explained to respondents and they were asked to provide their names and addresses so that a short questionnaire could be mailed to them. The interview was to be conducted by telephone; the purpose of the mailed questionnaire was to assist respondents in following the question sequence and comprehending the questions. During the initial phone call, respondents were also asked to specify a convenient time for the interview. More than one-third of the eligible respondents declined to give their names and addresses, thus depressing the final response rate and raising concerns about response bias. Most refusals occurred at the name and address solicitation stage; this approach is not recommended.

The interviewer's inability to control the response situation and the respondents' difficulty in consulting household records during a telephone interview are also disadvantages. When interviewing a respondent, we never know where the telephone is located, whether anyone else is in the room, and how comfortable the respondent is in answering the questions if someone else is nearby. Without advance notice, it is also very difficult for respondents to consult records when being interviewed by telephone.

Another disadvantage, which may not be obvious, concerns limited responses to open-ended questions. It might seem that a telephone interviewer could probe respondents' answers and record verbatim responses as well as a face-to-face interviewer. In fact, telephone respondents can and do respond to open-ended questions; however, they usually answer in a sentence or in a few short sentences. If the interviewer's probe for "other reasons," explanations, or clarification creates silence or "dead time" and because the respondent cannot think of more to say, the respondent may begin to feel anxious or edgy. Research shows that long and detailed answers are not elicited as often in telephone surveys as in face-to-face surveys (Groves & Kahn, 1979).

Finally, information about refusals and noncontacts is quite limited with telephone surveys unless the sampling frame includes names and addresses or other identifying information. If we have names and addresses, we can, for example, compare response rates by gender, area of the city, or other available characteristics. With RDD, this is only partially possible; we cannot be certain that a telephone number is that of a household, much less find out who lives there, unless contact is made or an interview is completed.

Face-to-Face Surveys

In face-to-face surveys, also referred to as personal interview surveys, information is usually collected by interviewers in the home or in another location that is convenient for the respondent. The key element is that respondent and interviewer are together in the same location.

This is the most expensive of the four methods because of the travel costs involved and the amount of time needed to collect the data. The interview itself takes only approximately 25% to 40% of that total; travel, the editing of responses, and other tasks consume the remaining 60% to 75% (Sudman, 1967). To understand the travel time required, assume we need to conduct 800 interviews with persons 18 years of age and older who reside in households in a large city such as Chicago. Assume that an average of five interviews should be conducted per selected block. If we want each interviewer to do an average of 25 interviews, we need 32 interviewers (800 ÷ 25 = 32). Each interviewer will be assigned 5 blocks to work (25 interviewers ÷5 interviews per block = 5 blocks). The city of Chicago is approximately 234 square miles and contains more than 10,000 blocks. Because blocks will be selected randomly, each interviewer may need to travel 1 mile, 5 miles, or even more to reach each of the interviewer's assigned blocks, and may need to return several times to contact all of the selected people. Clearly, travel consumes a major portion of the interviewers' total time.

Because the expense of face-to-face designs is greater than that incurred for the postage stamps used in mail surveys or for "letting your fingers do the walking," the main method of "travel" in telephone surveys, why would a researcher ever want to do a face-to-face survey? As we discuss shortly, it is the preferred method of data collection for some surveys because it is clearly superior for certain types of questionnaire items and for increasing data quality, especially when using CAPI.

Advantages

Although the face-to-face interview is the most expensive survey method, the cost disadvantage can be offset by many advantages. The cost is greater than that for the other methods because interviewers must travel to the respondents' homes and are typically paid for both their travel time and the interview time. The following brief description of how the sampling and interviewing are done for a national survey, based on the General Social Survey (GSS) model, illustrates why this method is expensive.

Assume that we want to do a national survey of adults 18 years of age and older who live in households in the 50 states. After considering existing lists, we conclude that no list or sampling frame contains the names and addresses of U.S. residents who are 18 years of age or older. For example, state motor vehicle departments only have information on people who have drivers' licenses; voter registration lists are incomplete because not everyone registers; income tax forms are strictly confidential, and not everyone completes a form each year; even the U.S. Postal Service does not have a complete list of addresses.

Although there is no list of people or households in the United States that is complete enough to serve as a sampling frame, we can use the 2000 U.S. Census of Population and Housing to determine the number of people and the number of households within various political and geographic areas. For example, we know fairly accurately how many people lived in the city of Chicago, or in Wake County, North Carolina, or in College Park, Maryland, in April 2000.

The GSS design is like a funnel. First, the larger areas are selected randomly, in stages, using census information. In the GSS, there are 100 selections made at the first stage of selection. The second stage of selection is categories or divisions of the first stage areas. In cities, we might use groups of blocks or census tracts (areas of approximately 2,500 to 8,000 people); in counties, it might be cities, townships, or blocks; and in rural areas, sections of land might be selected. The number of second-stage selections is 384 and is based on the population or housing units of the first stage of selection. The third stage of selection is usually part of a block, or enumeration district, or section of land. In the GSS, units for this stage of selection must have a minimum of 50 housing units.[5] These first three stages of selection are based only on the number of people or housing units within the delineated areas. After the third stage of selection, "listers" are sent to each of the selected areas; their job is to list every housing unit in the selected areas. After all households are listed, a random sample of households is selected.[6] At each, an interviewer randomly selects one person 18 years of age or older to interview. An average of 2.7 interviews are conducted per block or area (Davis, Smith, & Marsden, 2001). Do you know how you obtain an average of 2.7 interviews per block or area?[7]

There are a number of sampling and data-quality advantages to the face-to-face method. Response rates are usually higher than those for telephone interviews. One reason for this is that an advance letter can be sent to the respondent's household prior to the interviewer's visit. The letter is usually printed on the sponsor's letterhead and explains the reasons for the survey, the importance of interviewing each sample household, the use that will be made of the data, and the confidentiality of the respondent's answers. This letter legitimizes the interview and makes the task of gaining the respondent's cooperation somewhat easier. Another response rate advantage is that it is more difficult to refuse someone face-to-face than it is to hang up the telephone. There are, however, some drawbacks. Conducting interviews in high-crime areas presents safety problems. Attempting interviews in apartment buildings can also be a problem because of difficulty in gaining access. As is true of telephone interviews, cooperation rates are higher in small towns and cities and lower in large metropolitan areas.

Sampling frame bias is usually low in the face-to-face method. When census data are used as the sampling frame, all individuals in the population, theoretically, have a chance of being included in the sample. Census data become a problem as time passes and the information becomes dated. This is especially true in areas of high growth or decline. Where high growth or decline occurs, the researcher must rely on local government statistics, recent surveys done in the area, or local informants such as real estate agents or banks to help update the population or housing unit counts.

Response bias is also usually low in the face-to-face method. The rate of cooperation is roughly equal for all types of respondents. Since the data are collected by interviewers, the face-to-face method has many of the same advantages as a telephone survey. However, face-to-face surveys allow more control of the response situation than do telephone surveys. For example, if the interviewer believes that the respondent's answers may be affected by others in the room, the interviewer can suggest moving to another part of the home to ensure privacy. Rapport is also better in face-to-face interviews because the respondents get to see the person they are talking to. Interviewers are trained to be courteous and attentive and to build rapport with their respondents. The quality of recorded responses in face-to-face surveys is rated as very good because interviewers receive extensive training in asking the questions and recording the answers. In addition, their work is thoroughly checked by both supervisors and the coding staff. However, there is also potential for unwanted interviewer effects. The interviewer's appearance, manner, and facial expressions are, unlike telephone and other data-collection methods, all on display and can potentially affect respondents.

Many of the advantages of the face-to-face method relate to the questionnaire itself. The questionnaire can be more complex because it is administered by a trained interviewer. For the same reason, and because both interviewer and respondent are in the same location, complex tasks or questions can be asked. For example, assume we have a list of 25 behaviors that vary by degree of risk, such as bungee jumping, sky diving, riding a mountain bike, and so forth. These 25 behaviors are listed individually on 25 cards, and we want each respondent to put them into one of four piles. The piles are "would never do," "would consider doing," "have done at least once," and "have done more than once." This task is much easier for the respondent if an interviewer is in the same room explaining and illustrating the task. Face-to-face interviews also allow for the use of visual aids. For long questions or long-answer categories, the interviewer can hand the respondent a copy of the question or the answer categories and have the respondent follow along as the items are read, making it easier for the respondent to understand what is being asked. Interviewers also have control of the question order in face-to-face interviews.

The face-to-face survey method is also best for open-ended questions, because these surveys allow a more relaxed atmosphere and tempo than do telephone interviews. As a result, it is easier for the interviewer to probe for additional information, and respondents are not as uncomfortable with long pauses between answers because they can see what the interviewer is doing (Groves & Kahn, 1979).

A face-to-face interview can be longer than a telephone interview for several reasons: it takes place in the respondent's home; respondents do not have to hold a telephone receiver while they listen to the interviewer; "dead time" or long pauses are not a problem; and the interview questions and tasks can be more varied, longer, and interesting.

Another positive aspect of the face-to-face method is that, because the interview takes place in the respondent's home, the respondent has the opportunity to consult records. Still, the respondent may not be able to locate the appropriate information while the interviewer is present unless the respondent has been given advance notice that records will need to be consulted. In this regard, mail and Internet surveys may be more advantageous because respondents can locate and consult records at their convenience. However, with panel studies that use CAPI, information from prior interviews can be programmed into the questionnaire and used to check respondent's answers or to assist them with recall (Nicholls II, Baker, & Martin, 1997). Finally, the face-to-face method is as well suited as the other methods to the asking of nonthreatening questions.

Disadvantages

There are four major disadvantages to the face-to-face method. The first is cost. As we explained earlier, it is expensive to complete the listing of housing units and to compensate the interviewers for travel time and other expenses. A national face-to-face survey with interviews lasting 45 to 60 minutes probably costs at least $250 per completed interview. Although cost data comparing the four methods are not readily available, a national face-to-face study probably costs more than twice as much as a similar telephone survey.

Another obvious disadvantage is the amount of time it takes to complete a face-to-face survey. Groves and Kahn (1979) reported that the entire interviewing process for a face-to-face survey is 2.6 times longer than that for a similar telephone survey. Considering the widespread distribution of interviewers across the survey area, training time, the transfer of completed questionnaires and materials to and from the interviewers, and other logistical considerations, the data collection phase of a face-to-face survey could easily take three times longer than that of a similar telephone survey.

To achieve some cost advantages, researchers attempt more than one face-to-face interview per selected block. If one respondent is not home, the interviewer can try another sampled household. There is a penalty, however, for *clustering,* or doing more than one interview per block. People of similar characteristics (such as income, demographics, and attitudes) tend to live on the same blocks or in the same neighborhoods. Thus, when we sample more than one respondent per block, we collect similar information for some variables. Think about it. People with similar incomes and lifestyles tend to have similar values and behaviors. A respondent may not be exactly the same as his neighbors, but he is probably more similar to them than he is to people who live on different blocks or miles away. Thus, when we do more than one interview per block, we are not capturing all the diversity in the sample population. However, all residents on a block are not going to be exactly alike on all the variables for which we are collecting information. Sampling statisticians have shown that the optimum design has some clustering rather than none. As a compromise between cost and a lack of diversity on a block, most social science surveys try to average between three and eight completed interviews per block. The more similar people may be on an important variable, the fewer the number of interviews we would want to complete per block.

Another possible disadvantage with face-to-face interviews is the hesitancy of respondents to report very personal types of behavior. The expectation is that the more personal the method of data collection, the less likely respondents are to report sensitive behaviors. The findings, however, are mixed. A few studies have found lower rates of reporting in face-to-face interviews for depression (Aneshensel, Frerichs, Clark, & Yokopenic 1982), sexual behaviors (Czaja, 1987–1988), and alcohol consumption by women (Hochstim, 1967). On the other hand, Bradburn, Sudman, and Associates

(1979) found no significant differences by method for the reporting of declared bankruptcy or arrests for drunk driving, and Wiseman (1972) reported no differences for questions on abortion and birth control. More recently, audio computer-assisted self-administered interviews (ACASI)—whereby the respondent listens to questions through earphones and follows along by reading the question on a computer screen and then keys in or selects an answer—is equal to paper-and-pencil self-administered surveys for the reporting of sensitive information (Tourangeau et al., 2000).

Another related drawback is that respondents are more likely to provide socially desirable responses in a face-to-face interview. This is especially true in studies of racial attitudes, and the outcome appears to be affected by the racial characteristics of respondents and interviewers. Campbell (1981) and Hatchett and Schuman (1975–1976) found that answer patterns were different when the interviewer and respondent were of the same race. When the racial characteristics are different, the respondent is much more likely to give a socially desirable response. Sudman and Bradburn (1982) indicate a small effect by method of data collection, claiming that respondents are most likely to **overreport** socially desirable behavior in face-to-face interviews and least likely to do so in mail surveys.

Combinations of Methods

There is no requirement that a survey must use only one method. Sometimes, depending on the research question and an evaluation of the factors listed in Exhibit 3.1, the best approach is a combination of methods.

A few examples illustrate the possibilities. Assume we want to do a survey of the students at a university that will include questions about date rape and alcohol and drug use. Let us also assume we have limited funds to conduct this study. The first possibility we consider is an Internet survey, but we quickly decide against that because many students may be unwilling to participate because of concerns about the confidentiality of their responses to an online questionnaire. Another cost-effective means of surveying the students is to select a sample of classes and ask students in the sampled classes to complete a self-administered questionnaire. In the time it takes to complete one interview, we can obtain responses for an entire class. Assuming that most professors will allow us to use part of their lecture time to administer the questionnaire, this is a very efficient survey design. However, what do we do about students who are absent on the day of the class visit? If the number is large, the results of the survey may have a serious response bias. Thus, we need to include those missing from class, or at least a subsample of them, and **weight** their responses.

Because it is very unlikely the professors will let us come back a second time to interview the missing students, we need to ask them to complete the survey by mail, telephone, or in person. All of these methods will be more expensive than the classroom method. Therefore, we want to subsample these nonrespondents. How many students we interview and how we do it will depend on our time schedule and resources. The quickest way would be by telephone. What would you decide?

One of the authors used a combination of methods in a national survey of physicians that was mentioned briefly in Chapter 2. The original plan was to do a national telephone survey. Resources, time schedule, and survey questions indicated this was the optimum approach. In the pretest phase, the response rates were poor—less than 50%—even when there were more than 15 attempts to contact some physicians. To find an approach to improve the response rate, the researchers conducted two focus groups with physicians from the survey population. One important finding of the focus groups was that physicians (and other very busy professionals) need a method of responding that is compatible with their schedules. The suggestion was to let them pick the way in which they wished to be interviewed: by telephone or a self-administered mail questionnaire. This strategy worked. The response rate in the main study was approximately 40% higher than in any of the three pretests.

Another combination frequently used when multiple interviews are required in a household (of married couples, for example) is to hold an initial face-to-face interview with one member of the household and leave a self-administered questionnaire for the other household member to fill out and return by mail.

Offering mail survey respondents the option of completing an online instead of paper-and-pencil questionnaire is another mixed-mode possibility. Couper (2000), however, provides a cautionary note regarding this type of mixed mode design. Two examples he cites are the Census Bureau's Library Media Center survey of public and private schools and the 1999 Detroit Area Study. In both studies, questionnaires were initially mailed to the sampled schools/individuals who were also given the option of completing the survey via the Internet. In the Library Media Center survey, 1.4% of the public schools and less than 1% of the private schools who responded chose the Internet option and approximately 8.6% of the Detroit Area Study respondents did so. Clearly, the researcher needs to weigh the potential benefits versus the cost of offering an Internet option to mail survey respondents.

A final example of combining methods is an Internet–telephone survey. Assume that you wish to survey the faculty members at a college or university. If the university has a directory listing all faculty, their e-mail addresses, and their office and/or home telephone numbers, an Internet survey combined with telephone interviews with nonresponders to the Internet survey (or a sample of nonresponders) would be an efficient design if the survey topic, length of the questionnaire, and the like are suitable for an Internet survey.

The possibilities for combining survey methods are limited only by the imaginations of the researchers, tempered with a careful assessment of which methods are most appropriate for the survey topic and potential measurement errors that may result from people giving different answers depending on the mode of data collection. Dillman (2000) also provides a number of useful examples of combinations of methods of data collection.

Notes

1. There are many forms of computer-assisted interviewing (CAI): personal interviewing (CAPI), telephone interviewing (CATI), and self-administered questionnaire (CASI). Many government and professional survey organizations use CAI. Internet surveys are one form of self-administered electronic survey methods that also include e-mail and interactive voice response (IVR), also known as touchtone data entry (TDE). Our discussion of electronic survey methods is limited to Internet (Web) surveys because they are arguably the most widely used and most promising of these methods.

2. In an attempt to standardize the use of and reporting of survey final disposition codes, the American Association for Public Opinion Research has published a free document entitled *Standard Definitions: Final Dispositions of Case Codes and Outcome Rates for Surveys*. The report covers random-digit dialing (RDD) telephone surveys, in-person surveys, and mail surveys. It discusses various types of completion and contact rates, and how to report cases of unknown eligibility, ineligible cases and partial interviews. See http://www.aapor.org.

3. A number of field experiments in U.S. government surveys, typically face-to-face surveys, have found monetary incentives effective in raising response rates without aversely affecting data quality (Shettle & Mooney, 1999; Ezzati-Rice, White, Mosher, & Sanchez, 1995). Also, incentives have been found to be effective with some "hard-to-reach" groups, which are defined as ". . . (1) the economically disadvantaged; (2) the educationally disadvantaged; (3) minority populations; (4) adolescents, youth, and young adults; (5) drug users and those with special health problems; (6) frequently surveyed professional or elite populations; and (7) transients and persons who do not wish to be found for legal or other reasons" (Kulka, 1996, p. 276).

4. It is important to obtain information from the post office about respondents who move. Requesting an address-correction card may assist the researcher in determining if the respondent is no longer eligible because of geographic residence or if followup attempts should be made. The post office charges a fee for supplying a forwarding address.

5. More units than will be used in one survey are included in the final stage of selection because these areas are used for more than one survey. Conducting multiple surveys in the same areas allows the sampling and listing costs to be spread across studies. For each study, different housing units are selected.

6. One additional point should be noted. Interviewers should not live on the same blocks that are sampled because it is best if the interviewer does not know the respondent. When personal questions like those about household income are asked, most respondents are more comfortable telling a stranger than someone they know. Therefore, interviewers may have to travel 1 or more miles each time they attempt to obtain an interview with a sampled household, making this type of face-to-face method very expensive.

7. The same number of interviews are not completed on every block because of different rates of ineligible housing units (e.g., vacant units) and survey cooperation (e.g., refusals, unavailable residents). In an average of 10 blocks, 1 block may have 4 completed interviews, 4 blocks may each have 2 completed interviews, and 5 blocks may each have 3 completed interviews—a total of 27 interviews in 10 blocks, or 2.7 interviews per block.

4

Questionnaire Design: Writing the Questions

The survey questionnaire is the conduit through which information flows from the world of everyday behavior and opinion into the world of research and analysis; it is our link to the phenomena we wish to study. The focus of Chapters 4 to 6 is on how to transform research questions into serviceable questionnaires. Chapter 4 begins with an overview of the questionnaire design process. This is followed by two main sections: factors in questionnaire design and writing questions. Chapters 5 and 6 cover organizing the questionnaire and testing the questions.

Most, though not all, of our examples are drawn from two projects: a telephone survey of Maryland adults to determine their attitudes toward violent street crime, and a self-administered survey of University of Maryland students about several current issues on the College Park campus.[1] These two studies provide questionnaire examples for surveys of the general population and of a special population and allow us to note important differences between instruments that are interviewer-administered and those that are not. Yet, while there are a number of differences in both design and execution between telephone or in-person surveys and mail or Web surveys, the fundamental issues of questionnaire design for them are the same. We begin our discussion of questionnaire development with these basics. Some will seem simple, even mundane; but as we shall see, they are neither.

We must approach the task of developing a questionnaire not as an isolated effort but in the context of achieving our research goals within available resources. At the end of our task, we will have answered several questions and learned how to make a number of key decisions about the development of our instruments, including:

- Are respondents able to provide the information we want?
- How can we help them provide accurate information?
- What must respondents do to provide the information that is needed?
- How is each research question rewritten as one or more survey items?
- What questionnaire items should we include?
- Do the selected items measure the dimensions of interest?

Questionnaire Design as Process

The logical process of questionnaire construction, like other components of survey design and implementation, is a process that involves a planned series of steps, each of which requires particular skills, resources, and time, as well as decisions to be made.

Professional survey researchers develop questionnaires in diverse ways; and a particular researcher may even approach instrument design in alternative ways for different projects. It is possible, for example, to begin writing particular questions before the overall structure of the instrument is settled, perhaps starting with items of most interest or those that deal with especially difficult measurement. Alternatively, one might focus on one topic area among several that the survey will include if that area requires more new writing or has particularly difficult measurement objectives.

It is useful for those with less experience to follow a systematic process. We suggest a set of steps that are broadly applicable, while not claiming that this particular sequence of activities is superior to others, but simply that in our judgment it is one reasonable approach. The purpose of the process is to ensure that nothing important is overlooked and can also assist proper scheduling and budgeting. The steps in the process are listed below.

Questionnaire Development Steps

a. List the research questions
b. Under each research question list the survey question topics
c. List all required ancillary information (background variables, etc.)
d. Do a Web and literature search for questions from other surveys
e. Assess the variable list against the general plans for data analysis
f. Draft the survey introduction (or cover letter)
g. Draft new questions

h. Propose a question order
i. Revise "found" questions if necessary
j. Try out the draft instrument on a colleague
k. Begin revisions
l. Periodically "test" revisions on colleagues

Working through these steps systematically will help clarify the survey goals and, at the same time, begin to operationalize them. You can also begin to see the balance between writing new items, testing the instrument and other tasks.

Whether or not a particular project has formal research questions, there will at least be a set of topics of interest. List the topics and then break each one down into smaller descriptions of required information. This can be in the form of survey questions, but need not be. For the college student survey, we would list the issues of interest and under each issue list specifically what kinds of things we want to find out. Do we merely want to know the students' knowledge and opinions about, for example, university policy on alcohol and drugs, or also their personal behaviors. Do we want to know what they think should be changed about the current system of academic advising or just their experiences with it? List the research questions or topics, and under each list the relevant subtopics. Such a list might look like this:

University Policy on Alcohol and Drugs

Knowledge about current policy

Suggested changes in current policy

Behaviors of respondent

Reported behaviors of respondent's friends

Respondent's perception of the behaviors of other students

Academic Advising

Knowledge about the current system

Usefulness of the current system generally

Usefulness of the current system to the respondent

Experiences with the system

Suggested changes

In listing these possible topic areas, you can easily see what issues the survey results will address and which they will not.

Next, consider what ancillary information may be of interest. The basic issues are whether we are interested in making statements and obtaining estimates for our target population only as a whole, or whether we are concerned with subgroups of that population, whether to study the subgroups individually or to make comparisons between them. If we are concerned with subgroups, unless information to classify respondents is available from the sampling frame, we will need to obtain it in the survey. This includes simple demographic information that is often desired, such as age, sex, education, race, and/or income. We may know from other research that the answers to our substantive questions may be related to these demographic characteristics and other attributes (or we may want to find out whether such is the case). For the college student study, consider what ancillary information might be useful. If you want to do subgroup analyses to see if behaviors or opinions differ by student characteristics, for what attributes does it seem reasonable to expect differences?

Somewhat more difficult are other characteristics of the population that may aid our analysis. For example, in the crime survey would it be useful to know if the respondent or an acquaintance has ever personally been the victim of a crime? It may be that attitudes about sentencing, to take one example, are more highly correlated with respondents' attitudes than most demographics. Whether or not we are interested in this or other factors depends on our research goals and hypotheses. The point here is that if we fail to collect the information, some analyses will not be possible.

If serviceable questions are available from other surveys, a lot of work can possibly be avoided. If a topic is very specific to a population or issue, such as a campus issue, it is unlikely that useful questions can be found. However, on many topics, for example those related to crime victimization and the criminal justice system, many surveys, including large ongoing federal surveys, have addressed the issue.

It is worth a Web search by topic to determine whether appropriate questions are available. Of course, because a question has been used before does not mean it is flawless or even very good. It is a common error to accept survey questions uncritically because they "have been used lots of times" or "by experienced researchers for many years." Because a weak question has been used uncritically in the past is no reason for you to do so. "Experienced researchers" may be quite knowledgeable in their particular discipline, while being relative novices in survey design. With survey questions, as with any product, let the buyer (or, in this case, the borrower) beware.

We recommend specific consideration, during instrument development, of how the analysis will be done. Many researchers are very near finalizing their instrument before checking it against their analysis plans—if they ever do. However, it is also common that items are included in surveys that have no real analytic purpose, while other important items or background variables are either left out or not asked in a manner that is best suited to their ultimate use. Yet, this step is not particularly difficult.

Certainly, all analyses are seldom known beforehand. But thinking about what frequency distributions are of interest in themselves or, more important, how variables will be analyzed by demographic or other subgroups, or how variables might be used together in simple cross-tabulations can be very useful. It is not necessary to be exhaustive. Even setting up a few simple table shells or proposed charts or graphical presentations can be helpful, while requiring very little time.

Starting preliminary work on the introduction (or cover letter) is important because of cooperation issues (discussed in detail in Chapter 5). Additionally, it is a very useful exercise to write the succinct summary of the survey purpose that an introduction or letter requires. This step can force clarification in your mind of the essential survey purpose. It is imperative that the survey's purpose be made clear and concise to potential respondents; you cannot begin too soon to work on that goal. Conversely, you don't want to go into great detail with respondents. You want to provide just enough information to obtain cooperation. These steps have brought us to the point of drafting questions, putting them in order, and moving ahead with constructing and testing the questionnaire, the subjects of the following sections.

Factors in Questionnaire Development

In our examination of developing a questionnaire and writing questions, we focus on the response process, on the utility of individual questions, and on the structures of both individual survey questions and entire interviews. To this end, we continue with a brief overview of where the questionnaire fits into the survey research process and how the roles of researcher, interviewer, and respondent are related in the collection of the data.

When developing a questionnaire, we need to keep in mind the context and circumstances of the interview situation. Sudman and Bradburn (1982) call it a "conversation with a purpose." Schwarz (1996) reminds us that an interview is governed by Grice's "norms of conversation." These social factors also should be kept in mind when constructing the questionnaire if it will be administered by an interviewer, rather than filled out by the respondent.

Over the last 15 years or so, there has been extensive interdisciplinary research by cognitive psychologists and survey methodologists collaborating on issues of survey research design, particularly on the question–response process (Jabine, Straf, Tanur, & Tourangeau, 1984). While most aspects of this collaboration are well beyond the scope of this book, a model suggested for the overall survey process (Tourangeau & Rasinski, 1988; Presser & Blair, 1994), including respondent tasks, is a useful guide for our decision making in instrument development. By *model* we simply mean a formal description of how a process or phenomenon works. Models may be mathematical or, as in this case, diagrammatic. The purpose of a model is to describe the components of the process and to show how they operate together.

In this model of the survey process (see Exhibit 4.1), the researcher develops the survey question and specifies its analytic use. The question is administered by the interviewer or is self-administered. Then the respondent has to do four things: interpret the question, recall relevant information, decide on an answer, and report the answer to the interviewer. These answers are entered into a dataset, and the researcher analyzes the results, coming full circle back to the originally specified analytic use of the item. In designing the questionnaire, we need to keep in mind the roles of these players and their tasks, as well as the analysis plan, which is the analytic use to which the question will be put.

Before starting down the road of the procedures, techniques, research, folklore, and craft that go into questionnaire design, we need a sense of our destination. What are the characteristics of a good questionnaire? How do we know when we have reached our goal? The survey model will be our guide.

Exhibit 4.1   Model of the Survey Data Collection Process

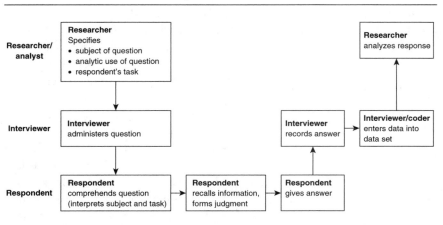

In this chapter, we note many characteristics of a good questionnaire, but three are so fundamental, they share equal billing at the top of our list: it is a valid measure of the factors of interest; it convinces respondents to cooperate; and it elicits acceptably accurate information.

When the researcher specifies the subject and analytic use of each question, clearly the underlying expectation is that a valid measurement will be obtained in the survey. Above all else, the instrument must measure the attitudes, behaviors, or attributes required by our research questions. Each research question may require one survey item or several. But if, in the end, we have not measured the factors of interest, along appropriate dimensions, our other efforts are useless. So the specification of the analytic use of each item has to be clear before we can write (or borrow) survey questions.

Assuming our questions are valid measures, respondents must then be convinced to cooperate and to provide acceptably accurate information. If we develop a questionnaire, however valid, that few respondents are willing to answer or, if willing, cannot answer reliably, we fail. Respondents must understand the question as the researcher intends, have the necessary information, and be able and willing to provide an answer in the form the question requires.

To accomplish these goals, we must not cling so tightly to the language of our hypotheses, constructs, or research concepts that few people other than experts can understand us. Neither can we be satisfied with a questionnaire that obtains cooperation and is easy to answer but at the price of distorting the measures we are after. Other factors contribute to a good questionnaire, but these three are paramount.

Suppose one of our research questions is, Do most people think that the criminal justice system is performing well? We might imagine a very crude first pass that simply tries to use this question as a survey item:

```
Would you say that most people think the criminal justice
system is performing well?

1. YES

2. NO

3. DON'T KNOW/NOT SURE
```

Assume that through pretesting, or just further reflection, the following criticisms of the question come up:

- Respondents cannot report very well about how people in general assess the criminal justice system.
- Respondents have different ideas about what is meant by *criminal justice system.*

- Respondents resist choosing between just "yes" and "no" options; they express feelings that fall between these alternatives.
- Respondents want to address specific aspects of the system, rather than judging it in general.

More might be said about this question, but these criticisms give an idea of the types of problems that come up. Although the question is exactly the research question of interest, it does not work as a survey item. Now consider the following revision, which addresses the three criticisms.

```
Do you think that people convicted of crimes receive fair
sentences almost always, most of the time, some of the time,
or hardly ever?
```

This item also has problems—as an exercise after reading this chapter, you should try to list what they are—but this modification does address the three criticisms of the original version: It asks what respondents themselves think, rather than asking for a judgment about the opinions of others. It eliminates the ambiguous and complex term *criminal justice system*. It provides a range of response categories to measure shades of opinion. But, in making the question more palatable to respondents, we have all but abandoned the research question: Do most people think that the *criminal justice system* is performing well? Why is that?

The new question fails in at least three ways, and possibly in others. First, it is too limiting. The criminal justice system involves more than sentencing, it also has arrest and rehabilitation functions, among others. Second, factors having nothing to do with the system, such as the financial resources of the accused, may determine how "fair" a sentence, if any, is received in particular cases. Third, the item does not get at the notion of a system, parts of which may work better than others. This research question is complex, and a large part of the survey could be devoted to it alone. If that is not an option, we might consider the following version, which has an introduction, multiple items, and open-ended followups:

```
Three important parts of the criminal justice system are the
police, the courts, and the prisons. I'd like to ask you
about the job each is doing.

  A. Do you think that the police are doing an excellent,
     good, fair, or poor job? {if fair or poor} What would
     you like to see them do better?
```

B. How about the courts? Are they doing an excellent, good, fair, or poor job? {if fair or poor} What would you like to see them do?

C. How about the prisons? Are they doing an excellent, good, fair, or poor job? {if fair or poor} What would you like to see them do?

In this version, we avoid the problems in the first question draft yet do not lose sight of the central research question. We define the term *criminal justice system* to help ensure that all respondents answer about the same things. We ask about the components separately; if people feel differently about some of them, we capture that. For those who do not think a particular part of the system works well, we find out why. While these multiple questions add complexity to our analysis, they also add richness and realism to it.

When operationalizing the research questions as survey items, we are forced to be concrete, to specify exactly what information we expect from respondents. This task may force us to reexamine some of our research questions. There is nothing wrong with this, providing we do it deliberately and consider the effects of such a reexamination on our analysis plan. At this key decision point, we need to ask the following questions: Exactly what data do we need? What will we be able to obtain? How do we plan to use it?

We turn now to the respondent's role in the model: comprehending the question, recalling relevant information, and forming and reporting an answer. First, note that respondents' comprehension requires interpreting both the question's subject and the task. For example, assume that, in the study of college students, we want to know how many courses in mathematics a college senior has taken since entering the university. We write a draft question: "How many college mathematics courses have you taken?" The respondent must understand not only the subject of the question, college mathematics courses, but must know exactly what to count (the task) in arriving at an answer. This may seem a simple task, but it isn't. Consider whether the following should be included in the courses counted:

1. Courses being taken now, but not yet completed

2. College-level mathematics courses, such as calculus, taken in high school

3. Remedial courses taken in college

4. Courses taken by transfer students at their previous college

5. A course taken but failed last semester

6. A course that primarily involves mathematics but is given by another department or is not listed as a math course, such as statistics

7. An audited course, not taken for credit

8. A course taken on a pass–fail option

9. A refresher or review course

It is often useful to construct such a list when a term might be commonly understood to include different things. As an exercise, try to write a new question that includes courses in categories 3, 4, and 8, but not the others.

Even this simple question on college mathematics courses requires a number of decisions by the researcher, decisions that affect how the question is finally written and what the results will mean. Thus, in writing questions, we must carefully consider exactly what we want respondents to do when we pose tasks such as counting or estimating. If the choice of what to count is left to the respondents, different people will include different things, introducing measurement error into the resulting data.

As the model indicates, the respondent must first comprehend the question, then recall relevant information. So many survey questions rely on the respondents' ability to recall information that it is useful to consider this area in some detail. In considering the role of memory in a survey interview, we tend to think of respondents remembering and reporting particular facts, events, or behaviors. However, we should also recognize that even opinion questions involve memory. If the respondent is asked for an opinion about something the respondent has thought about prior to the interview, then the response is simply a matter of recalling that previous judgment. If the respondent has not considered the subject before, or if something in the interview prompts additional consideration, then the respondent must recall information relevant to reaching an opinion. That new information is used to form the judgment, which is then reported (Tourangeau & Rasinski, 1988).

In developing a questionnaire, we are not concerned that some respondents have poorer memories than others, but that certain recall tasks are intrinsically more difficult. Respondents' particular past experiences, their general memory aptitude, and the cognitive effort and recall strategies they use all interact to affect their ability to remember information. In deciding what we can ask in an interview, we rely first on our judgment of how plausible we think it is that respondents can remember the requested information. Second, we try to confirm this judgment by pretesting the questionnaire.

Returning to our example of the college student survey, assume we want to know whether students think that there are adequate treatment programs on campus for alcohol and drug abuse. Some students will have thought about the issue or will have more knowledge about it, perhaps because of contact with such a program. They will respond easily with a previously formed judgment. Respondents who have not formed an opinion will need to think about such things as what they have read about such programs or, perhaps, about the experiences of people they know.

If, in another instance, we ask seniors in college to list the courses they took the previous semester, we expect all of them will be able to do so quite easily. What if we ask those same students to list the courses they took in the ninth grade? Although some will be able to do it and others will not, it is inherently more difficult for all of them. Many factors conspire to make the task harder: the passage of time; intervening events, such as other courses; and the low salience to college seniors of their ninth-grade courses. What if we now ask them to list the grades they received last semester? Again, we would expect that all of them could do it, although it might take slightly longer than remembering what courses they took. What about their grades in junior high school? Again this is a harder task for all, and we would anticipate differential performance. In the task of remembering grades, another factor—*uniformity*—comes in. For the student who got all As (or all Cs for that matter), the task is much easier because the student does not have to try to remember each individual course.

We see that while the memory abilities of respondents are a factor, so, too, are the task characteristics. We often know a good deal more about the latter. Survey researchers, in the course of designing a study, frequently discuss (with or without evidence from research studies) how difficult particular recall tasks are for respondents. What questions are included and how they are asked depend partly on that assessment.

We must also admit that it may not be possible to address every research question of interest through a survey. Sometimes a task will simply be too difficult for most respondents, resulting either in poor cooperation, inaccurate data, or both. For example, while it might be very interesting to an epidemiologist or a gerontologist to interview people age 65 years and older about the details of their adolescent health history, the low salience of many of the relevant events and the many years that have elapsed since their occurrence are likely to make these data too difficult to elicit in an interview.

In developing our questionnaire, we often use reasoning of the sort just described. This approach facilitates our decisions not only about what questions to include but also how to structure them to aid respondents' memory. For example, consider the question about mathematics courses. Rather than try to construct a single item, which presents respondents with a possibly difficult recall task, we might break up the task, asking either about courses semester by semester or about different types of math courses (e.g., regular, audit, pass–fail) in separate items.

The decisions we make will affect how well respondents can answer the questions we pose and whether they are willing to try. One sure way to obtain poor response rates is to ask questions that are difficult for many respondents.

Writing Questions

Now that we have at least a general sense of our goal of operationalizing our research questions into survey items, how do we begin to construct a questionnaire?

We can borrow questions other surveys have used or we can develop our own. Neither method is inherently better than the other. An advantage of borrowing questions from previous studies is that we can (or may need to) compare our results to previous findings. Such a comparison is much more problematic if our questions are different from those used previously. If our findings differ from the earlier study, we cannot say how much of the difference is because of the changed question. In addition, questions from other surveys may, depending on the former use, have already undergone a great deal of testing, saving us some effort.

One category of questions that it may indeed be important to borrow is that of screening questions, the items at the beginning of a questionnaire used to determine eligibility for the study. For example, assume we are looking for a particular characteristic, such as adults who were crime victims, and our estimate of the proportion of households with an eligible person is based on a previous survey. One of the first questions asked should determine if the potential respondent is a crime victim. If we ask the question about victimization differently than the previous study did, we may well find a different incidence of crime victims, affecting our screening rate and related costs. That is, if we expect a certain percentage of respondents to screen into the interview and, in fact, find a lower percentage, then either our costs increase to obtain the desired sample size of crime victims or we must settle for fewer such cases.

In general, we need to take care that borrowed questions truly measure the variables we intend. Using a question that is "close" to save time or effort can result in false savings when it comes time to analyze the results.

Even borrowed questions need to be pretested. Language usage and connotations change over time, and the context in our questionnaire is likely to differ from that of the question's previous use (Converse & Presser, 1986). The pretesting of borrowed items is also necessary if our administration mode differs from the original one. A question that worked well in a self-administered questionnaire may not be clear, or may have other problems, when heard over the telephone. Finally, we should note the population for which the original question was used. A question that was suitable for a survey of physicians may not work at all for a survey of patients, or for the general population. Despite these caveats, there are many very good survey questions available, and it would be foolish to ignore them. However, because borrowed questions must pass the same standards for inclusion in the study as do the questions we write ourselves, we discuss some of those standards next.

Criteria for Survey Questions

Within the limited resources we have to design and conduct our study, it may not be possible to include every question of potential interest in the study. Each question has a cost and must be justified. Exhibit 4.2 specifies several criteria that a question should satisfy before it is used in the study. This decision guide summarizes many of the issues discussed to this point and links them in a step-by-step procedure that can be applied to each question considered for inclusion in the study. Determining whether questions meet these criteria depends on judgment. Some survey questions may require pretesting before this determination can be made; for others, the answer will

---

Exhibit 4.2   Key Decision Guide: Question Utility

A. Does the survey question measure some aspect of one of the research questions?

B. Does the question provide information needed in conjunction with some other variable?
{*If no, to both A and B, drop the question. If yes to one or both, proceed.*}

C. Will most respondents understand the question and in the same way?
{*If no, revise or drop. If yes, proceed.*}

D. Will most respondents have the information to answer it?
{*If no, drop. If yes, proceed.*}

E. Will most respondents be willing to answer it?
{*If no, drop. If yes, proceed.*}

F. Is other information needed to analyze this question?
{*If no, proceed. If yes, proceed if the other information is available or can be gotten from the survey.*}

G. Should this question be asked of all respondents or of a subset?
{*If all, proceed. If a subset, proceed only if the subset is identifiable beforehand or through questions in the interview.*}

---

be obvious. This set of simple criteria may be consulted throughout the course of questionnaire development.

The first two points in the guide have to do with the need for the survey question; the remainder address the feasibility of obtaining the information. Clearly, if the item measures some facet of one of the research questions, we should consider including it. If the item itself does not address a research question but is needed to complement another item, we should also consider including it. For example, assume we asked question A: "Since living in this neighborhood, have you ever had to call the police?" and that those respondents who answered "yes" would be asked the followup question B: "How many times?" To compare respondents, an additional item would have to be asked, question C: "How long have you lived in this neighborhood?" Although item C may not be of interest in itself, it is needed to analyze item B.

The Structure of Survey Questions

A very wide variety of question types are available to the researcher, but relatively few types are needed in most surveys, and, on their surface, most of these are uncomplicated. But as we will see, many items that at first glance appear boringly simple pose serious problems for respondents, require difficult decisions for the researcher, and consume many resources for testing and revision, often more time and other resources than the novice researcher allocates to questionnaire development.

When writing questions, we must consider two components of our model of the data collection process: respondent and interviewer. We want to pose tasks that both players can perform reasonably well. Although we focus primarily on the respondents' tasks, we should keep in mind that interviewers have to be able to read the questions smoothly, pronounce the words clearly and unambiguously (especially in a telephone survey, when interviewers cannot rely on visual cues to convey that respondents are puzzled), and have clear instructions for recording the responses. To these considerations, we add a very important factor: We do not want to influence the respondent's answer in one direction or another. Consequently, as we develop our questions, we must constantly ask ourselves whether we have introduced bias into the data-gathering process.

We want to write questions that are unadorned and uncomplicated, as explicit and single-minded as a lawyer's interrogation. While we want survey questions to sound natural if administered by an interviewer or to read smoothly if self-administered, it is important to recognize that a survey question is a very special construct with a clearly focused purpose. As much as we labor to make them sound natural, survey questions are not simply forms of conversation. Ideally, survey questions are shorn of the vagueness, ambivalence, asides, and digressions of everyday speech. Conversely, although we take pains to ensure that our questions reflect our research goals, these questions do not typically contain the language of social science either. That is, words used in questions are simpler than the scientific intentions that drive them. Unlike either everyday conversation or scientific discourse, survey questions must function without supplementary explanation. Once the survey question is arrived at, we require that interviewers ask the question in exactly the same way each time, neither adding, deleting, nor changing words. While we often do not know why different wording of questions produces different results, that such effects occur is not disputed (Schuman & Presser, 1981). The reliability of the data obtained through survey research rests, in large part, on the uniform administration of questions and their uniform interpretation by respondents.

For those questions we choose to write from scratch, there are several options for how we articulate them, in what order we present them, the response dimensions, and the form of respondents' answers. We begin by defining some options, noting the generally preferred approach, then discuss why deviations from that approach are sometimes necessary or desirable.

Whether for telephone, face-to-face, or self-administered surveys, the questions can be broadly defined, open or closed. In the former, respondents are not given explicit answer choices; in the latter they are. For reasons that will become clear, we recommend using closed questions as much as possible. Data from open questions are essentially narratives that must be interpreted and coded; after the survey is over, the researcher is still a step away from having results that can be analyzed quantitatively. We will make some limited use of open questions, but our focus is on closed items.

A closed question has two parts: the statement of the question and the response categories. We need to devote equal attention to each part. We state the question as directly as possible, and we require that the answer be given in terms of one, and typically only one, of the answer choices provided. Often, specific qualifiers or conditions are part of the question. These added components restrict the meaning of the question to match our research objective. The simplest example is the yes–no response option. For example:

```
In the past year, has anything happened to you or your prop-
erty that you thought was a crime or an attempt at a crime?


   1. YES

   2. NO
```

While the choice of response options is relatively straightforward, even in this case we have made certain decisions, for example, not to include "don't know" or "not sure" options. Whether we include such options depends on our judgment about whether many respondents may not be able to answer simply "yes" or "no."

Imagine that a respondent returned home one day and found a broken first-floor window. Does it mean that someone attempted to break into her home, or was it an accident caused by children playing? However much thought she gives to the question, she may remain uncertain. Now consider a situation in which several months ago, a respondent's calculator was stolen at work when she left it unattended for a moment. It was of little value, and she saw no hope of getting it back, so she did not report the theft to the police or to her insurance company. When she hears our survey question, she cannot recall exactly when this rather minor theft occurred. Was it in the past year or not? Again, she may be uncertain.

For this question, then, it may be reasonable to add a "don't know" option. But these examples illustrate more. They show that we need to think through each question from the perspective of plausible situations to judge whether all or most respondents will be able to answer it in the expected manner. Often, thinking through a few hypothetical examples yourself or posing the question to a colleague will quickly reveal such problems. Such informal judgments and conversations are not substitutes for formal pretesting but a complement to it.

As an exercise, consider what kinds of difficulties some respondents might have with the following survey questions. Pay particular attention to the response categories.

BECAUSE OF CONCERNS ABOUT CRIME, how often do you avoid certain areas that you'd like to go to? Would you say:

1. Always

2. Frequently

3. Seldom

4. Never

5. Don't know [volunteered][2]

In the past year, would you say that the VIOLENT crime situation in YOUR NEIGHBORHOOD has gotten better, gotten worse, or stayed about the same?

1. Gotten better

2. Gotten worse

3. Stayed about the same

4. Don't know [volunteered]

The Use of Qualifiers in Survey Questions

As noted, many times we need to qualify or restrict a question in one or more ways. This is often done to make a respondent's task (such as recalling past events) easier, to simplify a complex phenomenon or judgment (such as rating different components of satisfaction), or to tailor the survey item to fit the research question (such as focusing on things that occurred in a particular time period). Here is a list (by no means exhaustive) of common components of survey questions and examples of them.

1. Specific reference period

Quite often research questions may concern events occurring only in a particular time frame, for example:

How often have you shopped at the Book Center *this semester, that is, since January?*

or

In the past year, did anything happen to you or your property that you thought was a crime or an attempt at a crime?

In most instances, the time period needs to be limited to make it possible for respondents to provide reasonably accurate counts of a behavior or an event. Deciding what time frame will work for most respondents is a combination of what we know has worked in other surveys and our judgment of the similarity of those studies to our own.

In other instances, we may be interested only in a particular time period for substantive reasons. If we want to compare attitudes about crime in the past year to data from previous years, clearly we must limit interview reports to the past year.

2. Summary judgment

Sometimes, to avoid the use of an excessive number of qualifiers or to recognize the complexity of an issue, we ask respondents for a summary judgment.

In general, would you say that crime in your neighborhood is
very serious, somewhat serious, not too serious, or not at
all serious?

Overall, how satisfied are you with your academic advisor?

Without these qualifiers, some respondents may balk at responding,
saying that some kinds of crime have gotten better and other types worse.
Similarly, some students will say that they are sometimes very satisfied with
their advisor and dissatisfied at other times. These qualifiers are intended to
elicit an "all-things-considered" type of judgment.

3. Adjectives and other restrictions

Just as we can limit reporting to a particular time, so, too, can we limit it to
a particular place (precisely or generally defined) or to one category of events.

In the past year, would you say that the *VIOLENT* crime situation in *YOUR NEIGHBORHOOD* has gotten better, gotten worse,
or stayed about the same?

This approach is often used when the research question calls for comparisons of different aspects of the same issues, for example, different types of
crime in different places.

4. Reasons for behaviors

If the research question or hypothesis posits reasons for behaviors (or, much
more difficult, opinions), then the question must attempt to elicit those reasons.

BECAUSE OF CONCERNS ABOUT CRIME, how often do you carry a
weapon for protection?

As an exercise, for the question, "How many times have you gone to a
physician?" write four versions: one that adds a reference period, one that
includes only visits for a particular reason, one that includes only certain
types of doctors, and one that does all three. To keep things interesting,
don't borrow from the examples given above.

Response Categories

Usually response categories are explicit. Their choice determines, to a great
extent, what our data will look like. Before considering some characteristics

of effective response categories, let us examine some of the kinds of things response categories measure.

Survey questions can measure an enormous range of issues: attitudes, such as favoring or opposing abortion, the effectiveness of local police, or rating job satisfaction; behaviors, such as frequency of moviegoing; personal attributes or facts, such as age, income, or race; and knowledge, such as how acquired immune deficiency syndrome (AIDS) is transmitted. How well such measures are obtained in a survey interview depends not only on how precisely the question is asked but also on how suitable the response categories are. Sometimes the appropriate categories follow naturally from the question. For example, in asking "What type of school did you attend just before coming to the University of Maryland?," we need simply to think of the possible places a student might have come from, such as:

```
High school
Transferred from a 2-year college
Transferred from a 4-year college
Returning after an absence
```

If we are uncertain that we have covered all the possibilities, a "Somewhere else (please specify)" category might be added. When the question concerns a fact or attribute, it is often clear what the response alternatives must be. In questions about opinions, the categories, even the correct dimension, may not be obvious. For example:

```
Depending on the circumstances of the case, some students
found responsible for the possession and/or use of illegal
drugs are given the option of random drug testing for a period
of 2 years in lieu of actual suspension from the university.
How satisfied are you with this policy? Very satisfied, some-
what satisfied, somewhat dissatisfied, or very dissatisfied?
```

We could focus on which "satisfaction" categories to use, but something more fundamental is amiss. Is *satisfaction* the word that we want? Perhaps students who have had experience with this policy could say how satisfied they were with the result, but that's a somewhat different issue. For other students, satisfaction with a policy they have never encountered is imprecise at best. Maybe what we really want to know is whether students think the policy is *fair or appropriate*.

In other instances, response categories are integral to the question and just have to be listed for coding. A simple example is, "Do you sell your books after completing the course or do you keep them?" The categories are obviously "sell them" and "keep them." But even in this relatively straightforward case, we need to be sure all the possibilities are covered. Examining this question more closely, we see that a student could also choose to "sell some and keep some"; so we need to add this option. Are there any other possibilities?

A more complex question with integral response categories comes from the crime study. Suppose we want to find out what respondents think about alternative sentences for criminals, that is, a sentence other than being sent to prison. We might develop a question such as

```
Do you think people convicted of robbery should be sent to
prison, required to report to a parole officer once a month,
required to report daily, or monitored electronically so
their whereabouts are known at ALL times?
```

Again, the response categories are the heart of the question and are developed with it. But many times, in asking about attitudes, ratings or evaluations, or behaviors we have quite a choice of ways to structure the response options. Exhibit 4.3 lists a number of common response categories for measuring attitudes, behaviors, and knowledge. While this exhibit is far from exhaustive, it serves as an introduction to response categories. (For more examples of scale and question construction, see Sudman & Bradburn, 1982.) Note that some of these categories are easily adapted to other dimensions. For example, the "very/ somewhat/not too/not at all" quantifiers work well with measures such as "helpful," "serious," "concerned," "likely," and "interested," among others.

It is important to be aware that many of these terms do not have an absolute meaning. They are ordered relative to each other, but their meaning is left to the respondents' judgment; one person's "good" is another person's "fair." (Also see Bradburn & Miles, 1979, on vague quantifiers.)

Most of the response categories listed in Exhibit 4.3 use three or four choices and do not present a neutral category. We recommend keeping the number of categories small, especially for telephone interviews. Respondents have to keep the categories in mind while deciding their answer. Similarly, many respondents may take the easy out with a neutral or "no opinion" category if they do not feel strongly about an issue. Many of these respondents will, however, lean one way or another, which can produce useful data. Still, if, after pretesting, we find respondents pushing for a middle alternative, providing one might be advisable. There are widely divergent views on the number of categories and the use of a neutral category, which are beyond the scope of this book (see Sudman & Bradburn, 1982, and Schuman & Presser, 1981, for a full discussion of the issues).

Exhibit 4.3   Some Common Response Category Quantifiers

Opinions

Completely satisfied/ Mostly satisfied/Somewhat satisfied/Dissatisfied/Very dissatisfied

Very important/Somewhat important/Not too important/Not at all important

Oppose/Support

Strongly oppose/Oppose/Support/Strongly support

Knowledge

Very familiar/Somewhat familiar/Not too familiar/Not at all familiar

True/False

A lot/Some/A little/Nothing

Frequency of Events or Behaviors

Never/Less than once a semester/Once a semester/Twice a semester/Three times a semester/More than three times a semester

Per day/Per week/Per month/Per year/Never

Always/Frequently/Seldom/Never

Always/Sometimes/Never

All/Most/Some/A few/None

Often/Sometimes/Rarely/Never

Ratings

Gotten better/Gotten worse/Stayed about the same

Excellent/Good/Fair/Poor

A great deal above average/Somewhat above average/Average/Somewhat below average/A great deal below average

Very fair/Fair/Unfair/Very unfair

High/Medium/Low

Small/Medium/Large

In another type of response format, only the end points of a scale are labeled, as in:

```
On a scale of 1 to 10, where 1 means not at all serious and
10 means extremely serious, how serious a problem is crime
on campus?
```

This approach elicits a finer partitioning of responses (variance). However, it does introduce complexities into the analysis. Does one compute average ratings; combine parts of the scale into high, medium, or low categories; or use some threshold level (such as any rating greater than 7) to indicate that crime is perceived as a problem? All these and others are possible. If there are many points on the scale, it will probably be somewhat easier for respondents if the instrument is self-administered and they can see the range.

We should recognize the subjective nature and the inherent complexity of opinion questions (see Turner & Martin, 1984, for a full treatment of these issues). When two student respondents independently answer the question "How satisfied were you with the book center's special-order service?," the word *satisfied* may well have somewhat different connotations for each of them. That is, the dimension of satisfaction has multiple aspects, which may be given greater or lesser emphasis by different people (or in different situations). For one student, the satisfaction rating may be dominated by a particular, unpleasant encounter with a cashier; for the other, the perception of average costs of goods may determine the rating. Each respondent understands the term *satisfaction*, but each puts his or her own spin on it.

Some additional lessons for questionnaire design and interviewing come from this observation. First, if obtaining a good understanding of a respondent's satisfaction is central to the study, as it is in this case, multiple questions are probably necessary. After obtaining the general measure, some followup items would be asked. In this example, we might follow with a question such as "What was the MAIN reason you were {satisfied/dissatisfied}?"

Second, respondents may want to know what aspect of satisfaction is intended. When they ask an interviewer, the standard response should be "Whatever it means to you," letting the respondent make the choice of what kinds of things to consider (Fowler & Mangione, 1990).[3] There is a great potential for interviewers to influence responses in such situations, especially if the respondent does not have a strong opinion about the particular issue. This is one more reason why we emphasize that interviewers read questions verbatim and refrain from interpretive "helpful" comments.

Finally, in seeing again the complexity lurking within another seemingly simple situation, we may choose to add to our pretest agenda methods to explore how respondents understand such terms as *satisfaction, serious,* and others in specific question situations. This need may also affect our choice of pretest methods. We return to this in some detail in Chapter 6.

In constructing categories for frequencies of occurrence of events or behaviors, our choices of both the number of categories and the number of times denoted by each one is guided by our perception of the range of times the event or behavior is likely to occur. If we overestimate or underestimate this frequency, most respondents will be grouped into one category and we learn nothing. In such a situation, we say the measure has little or no variance. The semester-based response scale in Exhibit 4.3 (under "Frequency of events or behaviors") may work quite well for a question about visits to an academic advisor or trips to the bookstore; it would probably not work for visits to the library. Presumably, most students will go to the library many more than three times a semester, so our responses would be bunched into that one category. The choice of response scales is still another instance in which the questionnaire is guided by our perception of how the world (in this case, the world of academia) works; by pretesting, we evaluate those perceptions.

Sometimes we don't have even an initial impression of the correct choices to provide. In such a case, we may choose to ask an open-ended question (without explicit response alternatives) in the pretest to learn what frequencies of behavior (or other categories) are likely. We then use this information to construct closed categories for the main study questionnaire. For example, if we are not sure how often, on average, students see an academic advisor each semester, the pretest question might be simply: "All together, how many times did you visit your academic advisor last semester?" After checking what respondents reported in the pretest, it may be possible to create categories such as "never," "once a semester," "twice a semester," "three times a semester," or "more than three times a semester." On the other hand, if the pretest data show that many students meet with an advisor many more than three times a semester, then a set of categories that take this information into account might be constructed.

If we are not confident, even after pretesting, what frequency range makes sense, it may be best to ask the question open-ended, even in the main study, and code the responses later.[4] This situation could arise, for example, in a question that only a subset of respondents are asked, in which case pretesting does not obtain enough information to develop categories.

Identifying Weaknesses in Survey Questions

As with composition in general, writing survey questions is largely a process of rewriting. To improve questions in subsequent versions, you first have to examine each question critically to spot possible flaws. A good aid to this process is the use of multiple readers, either independently or in a group meeting. Most questions you are likely to use are not written for some specialized population; the questions should be clear and make sense to the general reader. It is often the case that if five people carefully read a questionnaire, many of them will spot identical question problems; but each of the five is likely to spot some flaw or issue that the other four do not. Almost any questionnaire will benefit from multiple readers.

Of course, "flaws" may be identified that you, as the designer, do not think are flaws, but you now have a basis for discussing the possible problem with your "readers." Once you spot a question's weaknesses, you have taken an important step toward improving the question. You can then revise the questions to eliminate the flaws while maintaining the measurement objective of the question. It is not uncommon for particular items to be revised many times before arriving at a satisfactory version.

Effectively revising a question can be very difficult; but often, once a problem is uncovered, the item can be greatly improved, if not made perfect, by fairly simple changes. Even if you are not entirely satisfied with the revised version, you have made an important start. Of course, no matter how proficient you become, you will always miss some flaws. This is why pretesting is essential, even for the most experienced survey researcher.

Some question structures are inherently troublesome. One type of question we suggest avoiding is the agree–disagree format. In this type of question, respondents are given a statement and asked whether they agree or disagree with it. There are usually two, four, or five categories provided. The following is an example of the agree–disagree format:

```
I am going to read you some statements about the criminal
justice system. For each one, please tell me if you strongly
agree, agree, disagree, or strongly disagree.

The police are doing an excellent job. Would you strongly
agree, agree, disagree, or strongly disagree?
```

Although this form is quite popular, it is also problematic and has been strongly criticized by survey methodologists (Converse & Presser, 1986). Research shows that there is a tendency (called "acquiescence response set") toward agreement, regardless of the question's content. Furthermore, there is evidence (Schuman & Presser, 1981) that this tendency is related to education, with less-educated respondents showing the tendency toward agreement more frequently than better-educated respondents.

Another type of question to avoid is the double-barreled question which, often unintentionally, has two parts, each of which the respondents may feel differently about. For example, "Do you think the police and courts are doing an excellent, good, fair, or poor job?" A respondent who thinks the police are doing a poor job but the courts are doing a good one has no way to answer. Unlike the agree–disagree items, double-barreled questions do not reflect bias or the researcher's preference; rather, they are structurally flawed. While judgments of experts may differ on whether an agree–disagree item is weak, there is not such room for discussion about double-barreled questions; they are always unsound.

The solution is almost always to break up the item into two questions. Double-barreled items often result from a misguided attempt to save space by piling multiple topics into one question. The result is questions that are unintelligible and unanswerable and cannot be usefully analyzed.

Lastly, ambiguity is the ghost most difficult to exorcize from survey questions. In testing a question on respondents, we find that they arrive at different interpretations of it so often that it almost seems willful. If first we banish ambiguity, all our remaining work is easier. Payne (1951), in his classic work *The Art of Asking Questions,* ends with an excellent checklist of things to consider in writing questions, many of which have to do with ambiguity.

Notes

1. The telephone survey was actually conducted by the University of Maryland Survey Research Center; the student survey is a composite of several separate studies that were conducted with University of Maryland students.

2. Typically, "don't know" or "dk" responses are coded 8 (or 88), and refusals to respond are coded 9 (or 99). This allows the same code to be used for these outcomes throughout the questionnaire, regardless of the number of response categories, which simplifies data analysis. In the example shown here, the interviewer does not offer the "don't know" response as an option, but the respondent may volunteer it.

3. Notice that this is the opposite strategy from the one we take for factual questions about attributes or the occurrence of certain events or behaviors, where the "objects" of inquiry are well defined, such as car ownership, and we want to ensure uniform reporting across respondents.

4. Note that this type of open-end question elicits a single number, or a number of times per some time period, and so does not present the difficult coding issues of open-end narratives.

5

Questionnaire Design: Organizing the Questions

Questionnaires are typically organized into sections that follow the logic of the sampling plan, the data collection procedure, and question administration. Most questionnaires consist of an introduction to the interview, a respondent-selection procedure, substantive questions, background or demographic questions, and, possibly, a few postinterview items. While the form and methods for each of these components will vary by particular survey and by mode of administration, their purpose, as described below, generally remains the same.

Introduction. When respondents (or household informants) are contacted for the interview, they must be given enough information about the survey to induce their cooperation. This step is necessary even when there has been prior notification about the study.

Respondent selection. This step usually occurs in general population surveys when the unit of analysis is the individual. Typically, housing units are selected—in one or several stages—from the sampling frame; then, within each sampled household, an individual is randomly chosen for the interview. Omission of a careful respondent-selection procedure—in studies where the unit of analysis is the individual, not the household—can introduce serious bias into the sample; for example, the people most likely to answer the telephone or the door will be selected more often than other eligible members of the household.

Substantive questions. In this section, we ask questions to address each aspect of our research goals. This is the heart of the questionnaire, accounting for the majority of the data and, hence, the majority of our effort and costs.

Background questions. It is common practice in general population surveys to obtain some background information, usually demographic, about the respondents. There is no standard set of such questions, but sex, age, race, education, marital status, and income are frequently used. (See the crime survey questionnaire in Appendix B for a formatted set of several common demographic items.) There are three general reasons for obtaining these data. First, our analysis may require them. We may have hypothesized that they can help explain the variation in answers to our substantive questions. For example, if we think there may be a difference between the attitudes of whites and nonwhites on the effectiveness of the criminal justice system, we need to know each respondent's race. Second, we may want to compare the demographic distributions in our study to census data in order to assess the representativeness of our sample. Third, if we decide to use poststratification weights (see Chapters 7, 8, and 10 on sampling and report writing), these data will also be needed.

For surveys of special populations such as students, background questions may also be included, depending on the analysis needs. Our student omnibus survey in Appendix A, for example, might ask class rank, grade point average, and other such items. But we should have a clear idea how each item will be used before adding it, because background questions also have a cost as they add to the length of the interview.

Postinterview questions. After the interview proper, there may be a few additional questions for the respondent, the interviewer, or both. Such questions generally fall into two categories. First, we may want some information about the interview just completed. For example, we might ask the interviewer to rate how well respondents seemed to understand particular questions or whether respondents were reluctant to answer certain items. Respondents might similarly be asked whether there were any times when they were not sure what a question meant, or why they refused or could not answer items to which they did not respond. These sorts of scripted postinterview questions can be particularly useful on pretests of potentially difficult or sensitive instruments (see the discussion of pretesting below) but may also be included in some main data collection as well. We must, however, keep in mind that postinterview items add to the cost of the survey, so such questions cannot be idly included. They must pass the same tests of usefulness as any other questions.

Second, after the interview, in some cases, we may ask respondents for the names of secondary sources (relatives or others who will know the respondents' whereabouts should they move) if we plan to recontact the respondents or to send them a copy of the survey results.

Transitions and auxiliary information. In addition to these main parts of the questionnaire, we sometimes need to give respondents information before they answer a question, for example, descriptions of the two social service programs we want to ask their preference about. In other instances, we need to write transition statements that precede questions about a new topic.

One possible by-product of trying to provide helpful information, or transitions, is the introduction of bias. One must take care that the information given to respondents to assist them in answering a question does not bias their responses. To the general public, bias often suggests intentional efforts to affect answers. Certainly that can happen; but unintended effects are much more common. Bias is also a concern in writing transitions, which should be short, neutral statements to bridge the gap between sections of a questionnaire, alerting the respondent that the next questions are about a new topic.

Another function of transitions can be reassurance of confidentiality preceding a section that includes sensitive or threatening items, for example, about sexual behavior or some illegal activity. While confidentiality will have been guaranteed at the outset of the interview, it may help reduce item nonresponse to mention it again at a crucial point in the survey.

Some things to watch out for when providing information or transitions in a questionnaire are listed below. Each type of problem is followed by an example of a transition statement or section introduction.

- *Social desirability.* "The student health service is developing a new program to educate students about the dangers of alcohol. These next questions are about your drinking habits."

Questions about drinking are generally somewhat sensitive. It is not socially desirable to report drinking to excess. This transition would very likely add to that sensitivity. Respondent underreporting of the frequency and amounts of their drinking might well result.

- *Adjectives conveying positive or negative qualities.* "Over the past year, the student government advisors have worked hard to improve the referral service provided by academic advisors. The referral service is an integral part of the advising process. Part of the purpose of this survey is to find out how good a job students like yourself think advisors are doing in providing referrals."

Again, the transition and information may have the effect of encouraging positive evaluations in the questions that follow. Not all respondents will be so affected, but some, particularly those without strong prior opinions about the advising service, might be.

- *Links to authority.* "The president of the university and the board of trustees have proposed a plan to reorganize the academic advising system. Some groups oppose this plan. The next questions are about your opinions about the plan."

In this instance, the prestige and authority of important administrators are on one side, while on the other side are vague opposition groups. In what direction might some respondents be swayed in their assessment of the plan?

What are some ways to avoid the potential bias of transitions and background information? The simplest is to minimize their use. Often they may not really be needed. Second, avoid language that is likely to arouse strong feelings; interviews do not need to be colorful. Never indicate a position on an issue. If transitions and information are necessary, keep them brief, balanced, and bland. Finally, in the questionnaire-testing phase, don't forget that transitions and information are also part of the questionnaire and should be evaluated with the same scrutiny. Every part of the questionnaire has the potential to affect answers and needs the same attention given to the questions themselves.

Introducing the Study

The survey introduction serves multiple purposes. It provides a compact preface to the survey, telling the respondent the subject, purpose, sponsorship, and a few other details. It gives the prospective respondent sufficient information about the study to satisfy the needs of informed consent. Most of all, it elicits participation.

All of these objectives have to be done fairly quickly, especially in a telephone survey, where the respondent may already be predisposed to deny an unsolicited telephone request. Moreover, it is important to recognize that each word and phrase in the introduction can potentially affect all three objectives, and to use that fact to advantage. For example, the simple introductory sentence "We are doing a study for the state of Maryland's Summit on Violent Street Crime" conveys the survey's subject, sponsorship, and also suggests that the data will be used by the state, presumably to deal with issues related to crime.

One objective in designing the questionnaire is to convince potential respondents that the study is important enough for them to give their personal resources of time and effort. We want the respondent to take the study seriously and try to provide complete and accurate responses. Convincing respondents to do this is one of our first tasks, and it begins when we introduce the survey.

In the last several years, the issue of participation has become especially important. Response rates in general population surveys—particularly telephone studies—have declined seriously. While one may speculate about the reasons for the decline, there is little doubt that it has occurred. In planning and conducting a survey, the amount of time and attention given to all aspects of gaining participation should be considered on a par with getting the content right. It is of little value to have carefully developed other parts of the instrument if the response rate is so low that it undermines the credibility of the study.

Credibility is different from actual nonresponse bias or low survey reliability. When the response rates decline, the concern about nonresponse bias increases. However, we usually do not have a measure of that bias; it is a *potential* effect that may or may not actually exist. Some studies show that for particular cases low response rates can occur without serious nonresponse bias (see, e.g., Keeter, Miller, Kohut, Groves, & Presser, 2000; Curtin, Presser, & Singer, 2000). Still, in the absence of evidence to the contrary (which is typically the case), a low response rate will reduce confidence in the survey's results.

Maintaining survey participation has always been one issue among others; it is now the overarching concern in developing the introduction. Of similar importance is the interviewers' ability when introducing the study to answer respondent questions and address possible concerns. As you read the following advice on introducing the survey, try to think of questions each component of the introduction could raise in a potential respondent's mind, then consider how you would instruct an interviewer to respond to each question.

It is important to realize that there is a distinction between the survey topic per se and our particular study. It is not enough that the *topic* appeal to respondents; the *study* must also seem worthwhile. A respondent may, for example, be very interested in the problem of crime and think that it is an important societal issue. It does not follow, however, that the respondent thinks that our particular study about crime is important.

Although, for the sake of brevity, we will not repeatedly refer to "potential respondents," it is important to keep in mind that this is the case. Each person selected into the sample is a potential respondent; but that person is also a potential refuser or other type of noninterview.

Introducing the study to the respondent can begin with the cover letter sent with a mail questionnaire, with an advance letter preceding a telephone or personal visit study, or with the interviewer's introduction in a "cold" contact at the respondent's household. We discuss key aspects of each of these forms of initial contact in turn. First, we establish some guiding concerns.

As we consider how to develop introductions appropriate to each method of interview administration, we want to keep in mind the kinds of questions that respondents often have about a survey. These include

What is the study about?

Who is conducting it?

Who is the sponsor?

Why is the study important?

What will be done with the study results?

These are the typical things that respondents want to know before agreeing to participate in a study. The respondent's interest in these questions should seem self-evident, but many researchers overlook the common-sense basis of the respondent's decision to participate in the study. Other questions respondents may want answered include

Why is the study pertinent to me?

Why was I selected?

Some respondents may want to know even more about the survey, but there is a limit to how much information can be conveyed quickly in the introduction. The questions listed are those that come up in most surveys of the general population, and of many special populations as well. We may find that for a particular study additional information is essential. For example, if the study requires asking about sensitive or illegal behaviors, it will probably be necessary to give special assurances of confidentiality. If the sample was selected from a special list, such as patients in an HMO (health maintenance organization), it may be advisable to inform respondents where their names were obtained.

Telephone Introductions

How do we decide what information to include in the introduction to a telephone interview and how do we best convey it? The following is a draft of an introduction for a random-digit dialing (RDD) survey about crime in the state of Maryland:

Hello, my name is _____. I'm calling from the University of Maryland Survey Research Center. The Survey Research Center is conducting a survey for the governor's 1991 commission on violent street crime. The survey results will be used to help plan programs to reduce crime through prevention, education programs for young people, and better law enforcement. Your household was chosen through a random-selection process. Your participation is completely voluntary but is very important to the representatives of the survey. All your answers will be kept in strict confidence.

Take a moment to check whether all the questions that potential respondents typically want the answers to, listed earlier, have been addressed in this draft introduction. Then, before reading further, take a moment and try to imagine how *you* might react if you heard this introduction on the telephone. Remember, the respondent did not expect the call, and no advance letter was sent. All that is known about the survey comes from what is in this introduction.

All the key questions have been answered, but the introduction is far too long. The respondent picks up the phone and is asked to listen to a paragraph. Remember, we want to convey these key points, but quickly.

Here is a first pass at revision. Let's compare it to the original and analyze the reasons for the changes we have made.

Hello, I'm calling from the University of Maryland. My name is _____. We are doing a study about crime for the State of Maryland's Summit on Violent Street Crime. The results will be used to help plan ways to reduce crime. Your household was chosen through a random-selection process. Your participation is completely voluntary. All your answers will be kept in strict confidence. For this study, I need to speak with the adult in your household who is 18 or older and will have the NEXT birthday. Who would that be?

First, we mention *where* the call is from before giving the interviewer's name. Remember, many refusals come at the very outset of contact. If the first thing the respondent hears is an unfamiliar name, it could be a negative signal. Hearing the name of an organization, and one that sounds legitimate, alerts the respondent that this may be some type of business call, rather than a wrong number or a sales call. This is a small matter, but many of the decisions in designing the questionnaire are in themselves small, yet they have a cumulative effect.

Next, we substitute the word *study* for *survey*. There is some evidence (Dillman, 2000) that the word *survey* carries some negative connotations. Also, sales calls are often disguised as surveys, which many respondents are now very sensitive about. This issue may become more of a problem with the new National Do Not Call List. First, although the law distinguishes surveys from telemarketing, probably many respondents are not aware of that fact. Such respondents may think that your survey call violates the new law, making cooperation that much more difficult to obtain. Second, some unethical telemarketers already disguise their sales call as a survey; this may increase. It may even be that some telemarketers may ask a couple of survey questions they have no intention of using as a way to technically be in compliance with the law.

The details about the use of the study results have been cut back. There was too much of this in the first version. The mention of the governor was also removed. While mentioning the governor may have helped get many respondents to cooperate, others may have negative feelings toward the governor, which would lessen their likelihood of cooperation. Limiting the sponsorship to the state of Maryland should provide the broadest appeal.

Additionally, the information about how the respondent's household was selected and the assurance of confidentiality were deemed nonessential for this particular study. Even though we do not say how long the interview will take, if the respondent asks about interview length, the interviewer must give a truthful response. Interviewers should say how long the average interview is but that it may be somewhat shorter or longer for any particular person. Similarly, the other information we decided not to keep in the introduction should all be available to the interviewers to use as individual respondent situations warrant.

The within-household, random-respondent selection was incorporated into the revised introduction. This allows a smooth flow from providing information about the study into making the first request of the respondent. So we end the introduction with a subtle transition into the question-asking phase of the interview.

The final version of the introduction was shortened still further to the following:

```
Hello, I'm calling from the University of Maryland. My name
is _____. We are doing a study for the state of Maryland's
Summit on Violent Street Crime. I need to speak with the
adult in your household, who is 18 or older and will have
the next birthday. Who would that be?
```

Advance and Cover Letters

Advance or cover letters can be an important part of the survey. An advance letter is sometimes sent to respondents before they are contacted by the interviewer; advance contact can also be done by e-mail for a Web survey. A cover letter accompanies a mailed questionnaire. Although their focus is somewhat different, the purposes of advance and cover letters are quite similar, so we will treat them together. (See Dillman, 2000, for a complete treatment.)

In both cases, the purpose is to use the letter to help obtain cooperation. To this end, advance and cover letters provide information very much like that in interview introductions, but with crucial differences. First, unlike the interviewer-administered introduction, advance or cover letters must stand on their own. Second, letters are much more easily ignored than an interviewer. For these reasons, such letters must be eye-catching (yet professional), clear (but brief), and compelling (but neutral). The letter must stand out from the welter of junk mail most people receive and must speak for the researcher to the respondent, addressing the key obstacles to cooperation.

As suggested by Dillman (2000), these letters generally include the following content in the order specified:

- What the study is about; why it is important; how the study results will be used
- Why the respondent is important to the study
- How the respondent was selected
- Promise of confidentiality
- A phone number to call if the respondent has questions and who will be reached at that number

The cover letter should not exceed one page and it should be on the letterhead of the person signing the letter.

In mail surveys, the cover letter serves the function of introducing the survey, so it is advisable not to repeat this information in the instrument itself. In the questionnaire itself, respondents should be given a simple set of clear instructions for answering, and each section of the instrument should be labeled in a different typeface to clearly signal transitions.

What Questions Should the Questionnaire Begin With?

Having guided the respondent through the introduction, we are ready to ask the first questions. What guidelines do we want to follow to select those questions? There are two things to keep in mind: first, the respondent's decision to participate occurs in stages; and second, there may be a logical relationship between the questionnaire items or sections.

There is not a large body of research literature on what motivates respondents to participate in a survey (see Groves, 1989; Groves, Cialdini, & Couper, 1992; and Groves, Dillman, Eltinge, & Little, 2002, for a more detailed treatment). But we do have a lot of evidence that most refusals occur at the introduction or during the very first questions in interviewer-administered surveys. It may be helpful to think of the respondent as making an initial *tentative* decision to participate (or at least to keep listening) based on the information in the cover or advance letter and in the introduction.[1] Next, the respondent hears (or reads) the first question. Besides beginning to obtain data, what should that first question accomplish? It should have all the following characteristics:

- *Relevant to the central topic.* We do not want to tell the respondent the study is about crime and then ask a first question about how many years of education, for example, the person has completed. While this may be very relevant to the analysis (and will be asked later), many respondents will ask themselves (or the interviewer), "What does my education have to do with the issue of crime?" This is one reason we do not usually ask demographic questions first.
- *Easy to answer.* Most surveys include questions that vary in the degree of difficulty respondents have in answering them. It is best to begin with a question that most respondents will find relatively easy to answer. While some respondents may find a difficult question intriguing, many others will wonder what they've gotten themselves into, and may feel frustrated and, perhaps, terminate the interview. Remember that we want to get the cooperation of a wide range of respondents. As an exercise, after reading this chapter, go through the questionnaires in appendices A and B and note which questions *you* would find easy to answer, and which you would find difficult.
- *Interesting.* Little in the survey research literature speaks directly to the issue of interesting questions. Some investigations of interviewer-respondent rapport touch on the issue of interest indirectly, but they seldom address it as an independent factor. Yet, common sense should dictate that, other things being equal, anyone would prefer to answer an interesting question than a mundane one. In general, it is more interesting to give an opinion ("Do you think the amount of violent crime in your town is increasing, decreasing, or staying about the same?") than a fact ("Do you read a newspaper everyday?").
- *Applicable to and answerable by most respondents.* One way to begin an interview that definitely would not seem to hold respondents' interest is to ask a series of questions to which many people will answer "don't know" because the questions do not apply to them. A question such as "In general, how reasonable do you think the price of software is at the University Book Center?"

Many respondents will have to respond that they don't know because they've never purchased software.

- *Closed format.* While some respondents will enjoy the chance to answer an open-ended question that does not limit their response choices, many others will find this a difficult way to start an interview. This may be especially true if, prior to the interview, the respondent has not thought much about the issue at hand.

Although these guidelines help us in deciding what questions to begin with, sometimes we have to deviate from this ideal. It is important to know when this is the case and what to do when it happens. Perhaps the most common situation of necessary deviation occurs when screening is involved. When the survey target population is not identical to the sampling frame population, questions have to be asked at the outset of the interview to determine whether the household or individual fits the target population definition. For example,

```
How many children under the age of 18 live in this house-
hold?
_____ ENTER NUMBER AND PROCEED
(None) END INTERVIEW
```

One note of caution in asking "screener" questions like this is to not make it obvious what type of target households are eligible for the survey. Some respondents will answer one way with the intention of opting out of the study (e.g., a man with children answering "no" to the above question). A way to avoid this is to balance the question so it is not obvious what answer will terminate the interview (e.g., "Of the people who live in this household, how many are adults 18 or older and how many are children younger than 18?").

Many times the screening questions are rather innocuous. For example, when the target population comprises households in a particular geographic area, such as a county or school district, the screener will simply ask place of residence. In other cases, the questions may be relatively sensitive, for example, if only households in particular income categories are of interest. In a few cases, such questions could be so sensitive as to make their feasibility (without severe underreporting) doubtful, for example, in a survey of homosexuals.

Sudman and Bradburn (1982) note that sensitivity is often determined as much by the response as by the question. For example, a question about illegal drug use is not sensitive to a respondent who does not use drugs. Other items, however, such as questions about sexual behaviors, are sensitive to most people. In general, we recommend that the novice researcher steer clear of sensitive items, especially as screeners.[2] When mildly sensitive questions must be asked early, the following guidelines should be followed.

- Ask as few sensitive questions as absolutely necessary to establish the respondent's membership in the target population.
- Make clear to respondents why it is necessary to ask such questions.
- If costs permit, consider inserting an opening nonsensitive "buffer" question or two before the screening item(s) to establish some rapport with the respondent.

Finally, it may be necessary to deviate from the guidelines for the opening questions if sections of the questionnaire apply only to respondents with certain characteristics. For example, if the first section of the questionnaire is about people who smoke, then we must first ask respondents whether they smoke to know whether to ask the section about smoking.

Grouping Questions into Sections

Now that we have some notion about how to begin the questionnaire, we have to decide the order of the remaining substantive questions. There is little research on this issue, so we will again employ a set of guidelines. We begin with the same qualities of relevance, ease, interest, and available knowledge that we considered in choosing our first interview question. Because most terminations occur early in the interview, these same concerns should guide the choice of the first set of questions and may also direct the order of subsequent questions.

To these considerations we add internal logic and a smooth progression, or flow, through the questionnaire. Certainly, if some questions depend on prior responses, the order of these items is dictated by that logic. It is also useful for the respondent to sense the flow, or natural progression, of the instrument. This is worthwhile in both interviewer-administered and self-administered instruments. As Dillman (2000) points out for mail surveys, the respondent should have a sense of progressing smoothly and, one hopes, rapidly through the questionnaire. We want to avoid having questions on a particular topic scattered through the questionnaire. Also, if a set of questions uses the same scale, it is easier for respondents to answer them together.

Now we'll apply these guidelines to the crime survey cited survey. Exhibit 5.1 lists the survey topics. How might we arrange them in the questionnaire?

We want the first set of items to be easy. Certainly it is easier for respondents to answer questions about subjects they have thought about before than about those they haven't. Some respondents will have thought about

Exhibit 5.1   Topics for the Maryland Crime Survey

A. Background (demographic) questions

B. Personal experiences with crime

C. Opinions about the criminal justice system

D. Fear that they or their family will be victims of crime

E. Perceptions of the violent crime problem in their neighborhood and in the state

F. Perceptions of the nonviolent crime problem in their neighborhood and in the state

G. Things done to protect themselves from crime

H. Opinions about alternative sentencing

I. Perceptions of the crime problem, generally, in their neighborhood and in the state

the criminal justice system, but many others will not. Also, as we have already seen, that set of items may be moderately difficult to answer. Personal experiences with crime meet the criterion of ease, as do things respondents have done specifically to protect themselves against crime. Both these topics, however, have the drawback of being somewhat sensitive. Alternatively, given the media attention to the crime problem, many people will have perceptions about how serious crime is, and the topic is also clearly relevant to the survey's purpose. Thus, we might begin with the questions on perceptions of crime.

It is more likely that people have thought about the crime problem where they live than somewhere else. So it is probably easier to ask first about neighborhood crime, then crime in the state. Should we ask about violent crime first, then other types of crime, and crime in general last, or should we follow some other order? What guides our choice? Sometimes the answer to one question can influence the answer to a subsequent item; that is, responses can be affected by the context in which they are asked. These context effects are often difficult to anticipate but it is important to be aware that they can occur and to think about where this might happen.

For example, a general assessment can be affected by the context of having already given more detailed ratings. When providing the general assessment, respondents will sometimes not consider the relevant detailed answer they have already given. Sudman and Bradburn (1982) speak of this as a "redundancy effect" and recommend that "overall, then, if you are asking a series of questions, one of which is general and the others more specific, it would be better to ask the general question first" (p. 144). However, research indicates a more complicated situation. For example, if a single specific question precedes a general question, many respondents are likely to exclude the content of the specific question from their answer to the general question (Strack, 1992; Tourangeau, Rips, & Rasinski, 2000). When a number of specific questions precede a general question, respondents, on average, are more likely to summarize their responses to the specific questions when answering the general one (Schwarz, 1991).

Following this advice, we might first ask about crime in general (I), then about violent crime and nonviolent crimes (E, F). Because perceptions are strongly influenced by personal experience, it will probably be easy for respondents to recall any experiences with crime at this point in the interview. Keeping in mind the concerns about the role of respondents' memory in answering questions, this might be a good place in the interview to ask about any personal experiences with crime (B). A minority of respondents or their families (although more than you might guess) will have been victims of a crime, but this question should certainly trigger their concerns about being victims (perhaps by bringing to mind an incident they read about or something that happened to a friend). So the question about fear of crime seems natural to ask at this point (D). Then, because actions taken to protect themselves and their families would, in part at least, be based on perceptions, experiences, and expectations about crime, logic would seem to point to this section next (G).

Finally, we are left with the items about the criminal justice system and alternative sentencing. Which order do we choose here? We again have a situation of a general and a specific aspect of an issue. So, following the advice of Sudman and Bradburn (1982), we ask first about the criminal justice system in general (C) and then about sentencing, the specific part of that system (H). As we shall see, the alternative sentencing questions are also somewhat more difficult, reinforcing the argument for placing them last. We end, as is typical in telephone interviews, with the demographic items (A). The final order (at least until we pretest) is shown in Exhibit 5.2.

Having walked through one question-ordering problem, we turn to the survey of undergraduate students about a wide range of topics for different sponsors. Rather than being driven by a set of research questions, the focus of this study is to obtain information for planning purposes by different campus administrators. Such a multipurpose study is called an omnibus survey.

The questionnaire is designed to be self-administered in a sample of classrooms. Responses are anonymous, rather than confidential. Unlike a confidential survey, in which the researcher can link respondents by name, phone, or address to their interviews, the procedures of an anonymous survey prevent anyone, even the researcher, from linking respondents individually to their answers. This approach is sometimes used when there are questions

Exhibit 5.2   Order Determined for Topics in Maryland Crime Survey

Perceptions of the crime problem, generally, in their neighborhood and in the state

Perceptions of the violent crime problem in their neighborhood and in the state

Perceptions of the nonviolent crime problem in their neighborhood and in the state

Personal experiences with crime

Fear that they or their family will be victims of crime

Things done to protect themselves from crime

Opinions about the criminal justice system

Opinions about alternative sentencing

Demographic questions

Exhibit 5.3   Topics for the Omnibus Survey of Students

Use of the university book center, shopping habits, and satisfaction

Use of academic advising and opinions about the advising system

Demographics

Alcohol and drugs: use, opinions on campus policy, perceptions of needs for education and treatment services

Race relations on campus

about very sensitive issues or illegal behavior. Given the topics shown in Exhibit 5.3, decide what order they might be arranged in. Then see the questionnaire in Appendix A for an example of one way they might be ordered.

Questionnaire Length and Respondent Burden

Telephone interviews typically take from 10 to 20 minutes or so to administer, although some may take only 5 minutes and others as long as an hour. In-person interviews commonly run 30 to 60 minutes, although interviews of 2 hours or more are possible. Mail questionnaires tend to be four to eight pages, but, again, many are a bit shorter and some a great deal longer. Why all the variation? How long should a questionnaire be?

Like many of the decisions in survey design, interview length requires balancing several factors, often without a clear rule for how to do it. The money and other resources available and the amount and type of information needed are primary concerns. But the expected interest or motivation of respondents and the limitations of the mode of administration are also very important.

Regardless of the available resources or the amount of information we feel we need, respondents must be willing to spend the time required to complete the questionnaire. Putting aside the possibility of monetary compensation, what motivation do respondents have to devote time to our study? If the study is interesting, people are more likely to want to do it. If it seems an important societal topic, they may have additional motivation; and, if some action will be taken on the basis of the survey, that motivation may be increased (Dillman, 2000; Groves, 1989).

For mail and other self-administered surveys (e.g., e-mail or Web surveys), it is paramount that appearance be carefully considered. The appearance of length can be as important an influence on respondents as the actual number of pages or questions. Dillman (2000) gives the most complete treatment of self-administered questionnaire formatting. We will note only a few key aspects of such formatting, taking the perspective of the survey-response model.

In making the decision to participate in the survey or not, the respondent has only a general idea of what is in store. At this point, the *perception* of burden guides the respondent's decision to participate or not. Respondents can quickly see the layout of a self-administered questionnaire, including spacing, type size, and question length. An instrument that is closely spaced, with tiny print or long questions, will be forbidding to many respondents.

Respondents can also be quickly put off by long or complex instructions for answering questions. We want to require as little work as possible of respondents in providing the necessary information. Unclear instructions can also have the effect of increasing respondent errors or item nonresponse among those who do participate.

We suggest (based on the work of Dillman) the following guidelines for the formatting of self-administered questionnaires:

- Limit instruments to six to eight pages.
- Precode response categories by assigning a number to each possible answer for the respondent to circle.
- Space the categories so that it is easy to circle one response without touching an adjoining one; arrange the categories vertically, one under another, rather than horizontally spread across the page.
- Provide simple instructions of no more than two sentences describing how to answer questions; for example, "Please circle the number of your answer unless otherwise instructed."

- Use a different typeface for questions, response categories, and transitions or section headings.
- Whenever possible, use arrows to indicate skip instructions. See Appendices A and B.

Avoiding Flaws in Mail Questionnaire Design

Two inadvisable design tendencies in self-administered questionnaires are so common that they deserve special mention. First, as in questionnaire design in general, we recommend minimizing the use of open-ended questions. Although these questions are easy to write and seem natural in an instrument that the respondent is going to write on, they present many problems. In addition to those discussed earlier, there are the added problems of illegible handwriting, unclear abbreviations, and ambiguous answers (which an interviewer is not available to probe). These can combine to cause serious problems in coding and analysis.

Second, inappropriately complex terms, whether jargon or "ten-dollar words," often find their way into self-administered questionnaires, perhaps because self-administered instruments are often (especially for mail surveys) pretested without the questions being discussed with respondents. Additionally, such surveys are often used with a special, rather than the general, population. We suggest putting the language used in self-administered questionnaires to a reality check by asking:

- Is the word found in everyday use, such as in the general section of a newspaper?
- Is there a simpler word that conveys essentially the same thing?
- If the term is purported to be the specialized language of a professional group, has it been checked with some members of that group? (Often, especially outside the physical sciences, even technical terms are not uniformly understood or used by all.)

Finally, the pretesting method(s) selected for a self-administered questionnaire should allow for feedback from respondents. Obtaining completed questionnaires, even those that look reasonable, is no guarantee that the instrument worked as intended. Respondents will often guess or take a stab at answering questions they don't fully understand. In our question about university drug policy, the term *in lieu of* may be known to most student respondents, but a nontrivial number may be unsure of its meaning and answer the question anyway. Without using a pretest method that obtains respondent feedback, this situation may pass unnoticed.

Web surveys also benefit from the same types of feedback from respondents. In addition, it can be useful to observe a few respondents answering a Web survey. There are issues of navigating through the instrument, locating information or instructions needed to complete the survey, and other "usability" concerns that do not apply to traditional, paper, self-administered instruments. These are discussed in more detail in Chapter 6, which discusses pretesting.

Notes

1. When asked, at the end of a telephone interview, the main reason they chose to participate, a nontrivial number of respondents reported being caught unaware and beginning the interview without really thinking about it. In those cases, it was really the absence of a decision not to participate (Blair & Chun, 1992).

2. For those researchers who do need to ask about sensitive behaviors, there are a number of factors to consider. For a recent, brief, but thorough discussion of the main issues, see Tourangeau and Smith (1996).

6

Questionnaire Design: Testing the Questions

The pretest is a set of procedures used to determine whether the questionnaire works in the manner intended by the researcher, providing valid and reliable measures of the attitudes, behaviors, and attributes of interest. Pretesting should, at some point involve trying out the questionnaire on some respondents selected from those who will be interviewed in the main study. Although there are some pretesting methods that do not use respondents, these methods should be used in conjunction with respondent-based procedures.

Everyone agrees that pretesting is essential, but it is useful to examine, in some detail, why this is so. This examination will provide a better understanding of survey questionnaires and also help us choose wisely from the array of available pretesting methods.

In designing a questionnaire, we make many decisions and assumptions, some conscious, others not. Underlying the questionnaire draft is our judgment about what kinds of things respondents will know, what words they will understand, what sorts of information they can and will provide, and what response tasks they can perform. When we pose alternative choices to the respondent, we have in mind some notion of the appropriate dimensions for an answer; in fact, we start to envision what our data will look like. Much of our effort is subtly informed by our feelings about how people will respond to our queries, by a belief that what we ask is sensible, by some vision of how the respondents' world works.

In a first draft of a survey question, some of these assumptions are likely to be wrong for many respondents. For example, do most people know how many total miles they drive each week? Will the typical respondent understand the word *export*? Can most people remember how many times they went to a restaurant in the past 6 months? All of these represent possible items we might use in a survey. Clearly, respondents differ in their knowledge, memory, and willingness to try to answer even moderately difficult questions.

In writing the first draft of the questionnaire, we have assumed that the vast majority of respondents will be able and willing to do the things the survey requires of them—our results depend on this being true. It is difficult, especially for novice researchers, not to use themselves as typical respondents, albeit often unintentionally, when writing questionnaires. In so doing, they assume that respondents can do the things they can do, understand the words they know, are willing to work as hard as they would to answer questions they feel are important or interesting. It is quite natural to make these assumptions, and just as natural not to realize they have been made. Whenever such assumptions can be identified and critically examined, the questionnaire is likely to benefit.

After the first pretest, we have data to assess some, though usually not all, of these assumptions. Many a questionnaire composed in cool solitude melts under the heat of real respondents' answers in the pretest.

Importance of Respondents' Comprehension of and Ability to Answer Questions

One thing we want to assess in the pretest is the respondents' ability to provide good answers to our questions. By "good" answers, we mean those that are valid and reliable. Validity requires, first, that the questions measure the dimension or construct of interest; and second, that respondents interpret the question as intended. Reliability mainly refers to the degree of variation in responses in repeated trials. For example, if another researcher uses our questionnaire on a sample survey of the same population (at the same time our study is conducted), following all our procedures, that researcher should obtain (within sampling error) the same results.

There are a number of pretesting methodologies available to determine how respondents understand survey questions. Although there has been some research comparing these methods (Bischoping, 1989; Presser & Blair, 1994; Cannell, Fowler, & Marquis, 1968; Campanelli, Rothgeb, Esposito, & Polivka, 1991; Willis, Trunzo, & Strussman, 1992), there are no definitive answers about which method is best in a particular circumstance. Once again, we must develop our own judgment and make use of general guidelines on how to proceed in best allocating our resources.

The methods for determining respondents' comprehension vary greatly by mode of administration. We first discuss pretesting when data collection is interviewer-administered (such as by telephone or face to face); then consider methods for pretesting self-administered instruments (such as by mail or on the Web).

The key resources in telephone survey pretesting are mainly, but not exclusively, labor (primarily interviewer hours) and available time. Other resources, such as materials, telephone charges, and the investigator's time to analyze the results, are needed, but the total costs are mainly driven by the cost of data collection staff labor. Pretesting is a nontrivial budget item and should be carefully considered during project planning. Failure to allow sufficient time and money for careful pretesting is a serious error.

Conventional Pretests and Interviewer Debriefings

In a conventional pretest, a small number of respondents, usually 20 to 40, are interviewed. The sample should be large enough to include people from diverse subgroups in the target population so that the questions and the answer categories are given a reasonable test. The exact procedures planned for the main study are followed. After this interviewing takes place, a debriefing meeting is held with the interviewers. This practice is so common that it is what most people generically think of as pretesting. But it is only one of several available methods (Cannell, Oksenberg, Fowler, Kalton, & Bischoping, 1989).

Bischoping (1989) notes differences of opinion about the ideal group of interviewers for the pretests. Some researchers favor using only the most experienced interviewers (Fowler, 1984; Converse & Presser, 1986); others suggest a mix of interviewer experience (Demaio, 1983) on the belief that less-experienced interviewers may uncover problems that the more experienced staff handle almost automatically. We suggest the former course. In a controlled experiment, Bischoping (1989) did not find significant differences between interviewers of varying levels of experience in the number or kinds of problems they mentioned.

Prior to pretest interviewing, it is useful to do two things. First, the interviewers should be trained just as you plan to do for the main study. In particular, any question-by-question instructions (Q-by-Qs) should be provided to the interviewers. If you want interviewers to answer respondent questions about the meaning of a term, for example, "health care professional" or "criminal justice system," this needs to be tried out in the pretest. These Q-by-Qs will affect respondents' answers. Of course, it is possible that in a small pretest no respondent will ask about a potentially troublesome term. For example, if only a small proportion of respondents are unsure of the meaning (in the interview's context) of "health care professional," a small pretest may not detect this. (This is one reason multiple pretest methods can be useful.)

Second, if the interviewers have not done pretesting before, it is a good idea to spend a little time discussing the purpose of the pretest and what kinds of information you want the interviewers to note. If they know you are interested in whether a respondent asks what a term means or if a respondent can keep response categories in mind, it will help the interviewers be alert to relevant respondent behaviors and to bring more useful information to the debriefing. Instructing the interviewers to make notes on a paper copy of the questionnaire after each pretest interview is a simple way to get richer, more accurate comments at the debriefing.

But a word of caution: You can inadvertently "plant" problems in interviewers' minds which they may well dutifully repeat in the debriefing. If you tell them, "I think some people are going to have real problems deciding 'how likely is it that [someone in their] household will be robbed or mugged,'" interviewers may overreport how often respondents struggled in answering that question. It is even possible that in anticipation of problems some interviewers don't give respondents enough time to decide on an answer or are more willing to accept a "don't know" response, thus affecting some respondents' answers—and your data. So it is advisable to be relatively generic in your instructions to interviewers about the kinds of things they should note.

The interviewer debriefing is generally organized as follows:

- *Overview of the pretest,* including any serious problems of respondent resistance to participation or to particular interview topics.
- *Question-by-question problem identification,* in which each interviewer is asked, in turn, about any problems the respondents had with the item.
- *Question-by-question suggestions for revision,* in which interviewers may suggest alternatives for handling identified problems.
- *Summary comments* on how well or badly the pretest went and on the main issues to be addressed before actual data collection or additional pretesting commences.

Note that even though we are interested in respondents' comprehension (as well as other respondent issues), the debriefing consists of a series of reports and opinions of *interviewers*. It is important to keep in mind that, in effect, the interviewer serves as a proxy for the pretest respondents. During the debriefing, it is often useful to separate interviewers' perceptions of *potential* problems that respondents might have from *actual* problems respondents encountered.

Additionally, in a debriefing, it is important to get comments from all the interviewers about their respondents' experiences and reactions. Debriefings can often be dominated by one or two talkative interviewers, while others with useful information or comments never speak up. One way to deal with this is to ask each interviewer, in turn, to comment on each questionnaire item.

Another problem to be alert for is what we might call the aberrant anecdote. Extreme interview situations are vividly remembered and almost always reported in the debriefing. Two common examples involve the extremely knowledgeable respondent and the outlier respondent. In the first case, say for our crime survey, an interviewer might have interviewed a lawyer who takes technical issue with many of the survey questions, points out alternative response categories, and raises many issues not included in the questionnaire. In the second, for the same study, an interviewer might encounter a respondent who recounts a large number of various crimes committed against household members, resulting in an unusually long interview and, perhaps, a suspicious reluctance to answer questions about household members' behaviors and precautions concerning crime. Such cases are likely to be raised at length in the debriefing, especially if they happened to one of the talkative interviewers. While we learn some things from these two respondents, we must keep in mind that they are atypical and must not let them dominate our thoughts on questionnaire revisions. Most respondents are neither legal experts nor frequent victims of crimes.

On the other hand, pretests often reveal instances of respondents' resistance to questions they believe are unrealistic. Consider our question about alternative sentencing:

```
Do you think people convicted of robbery should be sent to
prison, required to report to a parole officer once a month,
required to report daily, or monitored electronically so their
whereabouts are known at ALL times?
```

While many, even most, respondents may have been willing to answer the question in this form, others might have insisted that their answer depends on the circumstances. They want to know if a gun was used, if anyone was hurt, if the robber had prior convictions, and so forth. We might decide to go ahead with the general form of the question. But if this topic is central to the study and we know, perhaps from other research or from pilot work, the key factors that affect respondents' choices, we may choose to develop a more complex form of the item. In this case, we might provide a vignette[1] before asking the opinion alternatives. For example,

```
Suppose a [20]-year-old is convicted of robbing someone of
$100 on the street. He threatened the victim with [a gun].
He has [4] prior convictions.
```

We then ask the respondent's opinion about what should happen to the person. The items in brackets are varied across interviews following a pattern to produce a dataset in which we can analyze how opinions vary under different circumstances. Clearly, this question is much more complex to administer and to analyze than the version with which we began; it is also more realistic and informative. We learn more about the issue at hand, but, as always, at a cost.

Conventional pretesting is a powerful tool but it can be difficult to use well. We have noted some of the problems that can occur in the debriefing. The debriefing moderator needs to be well prepared and skillful in managing the group. Also, sometimes there will be conflicting findings, for example, some interviewers may think a question is problematic and needs revision, while others are just as convinced that it works fine as is. Sometimes a more experienced or articulate interviewer may be very convincing even when the weight of actual pretest experience does not support his position. Conventional pretests are good at suggesting possible problems, less good at providing clues to solutions.

Postinterview Interviews

Another common method of obtaining information about comprehension is through postinterview discussions with respondents about the interview they just did. This method is based on the work of Belson (1981), in which respondents to a survey were recontacted by a different interviewer and asked questions *about* the meaning of items in the first interview. Belson found that many respondents did not understand the original interview questions as intended. The current practice is to ask the postinterview questions immediately after the interview proper is finished. Because the respondent has just given time to do the main interview, try to keep the postinterview session brief.

The interviewer says something like, "Now, I'd like to ask you about the interview we just completed." Typically, we will want to use prepared postinterview questions but not limit the discussion to them. If it is clear from the interview that certain items caused problems or confusion, the interviewer should be given the latitude to include those issues as well.

Here is an example of a postinterview script:

Those are all the questions I have about the issue of crime in the state. Now, I'd like to ask you just a few questions about the interview.

1. First, were there any questions you were not sure how to answer?

   If YES, which ones were those?

   Why were you not sure how to answer the question about [subject of question]?

2. When I used the term *criminal justice system*, what did you think I meant by that?

3. When I asked the question about the quality of your neighborhood, what sorts of things did you consider?

4. Are there any questions you think that many people would find difficult to answer?

   If YES, which ones were those?

   Why do you think people would have difficulty with the question about [subject of question]?

5. Were there any important things related to these issues that we failed to cover?

Notice that, unlike the questions in the questionnaire, the items in the postinterview script are largely open-ended. Because we do not know exactly what types of problems respondents might identify, we do not want to overly constrain their answers. If this were part of a series of pretests, it may be that issues identified in the first pretest might be explicitly asked about in the subsequent ones.

For example, early postinterviews on the student omnibus survey might reveal such things as the importance to students of the availability of walk-in advising and their belief that peer advisors should be of at least the same class rank as the students they advise. These are the sorts of things that the researcher, not being a member of the survey population, might not be aware of, but such opinions can be uncovered quickly by allowing pretest respondents to comment on the study issues. (This is also a strength of focus groups.)

There are two ways to involve respondents in this sort of discussion. One way is to tell them at the outset of the interview that afterward we will be discussing the interview with them. Converse and Presser (1986) call this a *participating pretest*. The other is to conduct the interview as planned, then ask the postinterview items. This is called an *undeclared pretest*. Converse and Presser recommend beginning multiple pretests with a participating pretest, then moving on to an undeclared one. But there is little in the literature that addresses how well these two pretest methods work in combination.

Behavior Coding

Behavior, or interaction, coding is a method developed by Cannell et al. (1968) at the University of Michigan to quantitatively assess how well a face-to-face or telephone survey interview is conducted. In this procedure, it is assumed that in the ideal interaction between an interviewer and respondent, the interviewer reads the question verbatim, and the respondent replies with an acceptable answer. They then proceed to the next question. In interaction coding, a third party uses a simple set of codes to indicate deviations from this model: nonverbatim reading of the question, respondents' requests for clarification, requests to reread the question, or inadequate answers. Each time each question is administered, each problem type is coded 0 if it does not occur and 1 if it does. Then the percentage of times a problem occurs for each question is calculated.

Although this method was originally developed to investigate interviewers' performance, it has also been used to identify problem questions. Fowler (1989) recommends that questions that are coded 15% or more of the time into one of the deviation categories be considered problematic. While the 15% cutoff is somewhat arbitrary, it has proved useful in practice. An advantage of this procedure is that it can be done in conjunction with conventional telephone pretesting, at little additional cost, using monitoring facilities in a centralized telephone facility. If such a facility is not available, then added cost does become a factor. For personal interviews, tape recording and later coding is typically used. A commonly used set of behavior codes is given in Exhibit 6.1.

What do deviations from the ideal interaction tell us about the questionnaire? Interaction problems can arise because of interviewer performance or difficulties, because of respondent problems or because of problems in the question itself. It is not always clear what causes a particular problem. When in doubt do we lay the blame on the questionnaire or not?

Exhibit 6.1   Standard Behavior Codes

Interviewer

E(xact)

S(light)

V(erification)

WV (wrong verification)

M(ajor)

Respondent

AA (adequate answer)

IN (interruption)

QA (qualified answer)

CL (requests clarification)

IA (inadequate answer)

DK (don't know)

RE (refusal)

Deviation from exact reading can result from a question being awkwardly worded or overly long. This assumes that the interviewer tries to read the question as written, but has difficulty because of some feature of the question. But such deviations can also result from interviewers purposely changing wording.

Minor changes in reading are those that do not change the question's meaning at all. This can include such things as adding "filler" words or phrases, such as ". . . the next question asks" (followed by exact reading of the item). Changes such as substituting contractions would also be considered minor. Changing a word omitting or adding words are usually major changes, as is altering a question's syntax.

From the respondent's side, an interruption can mean that the question is worded in a way that misleads respondents to think that the question has ended. A respondent may also interrupt because he knows (or thinks he knows) the answer based on hearing part of the question. These types of interruptions occur simply because the respondent answers too soon. But interruptions can also occur because respondents are confused by the question. Interruptions can be caused by fairly minor structural problems in the question, but can also suggest something more serious.

When respondents qualify their answers it can mean that the question is unclear and the respondent does not know if he is understanding it correctly. But respondents will sometimes qualify their answers when the response task is difficult, leading them to be unsure of the correctness of their response. This often happens when a question requires a difficult memory task. An inadequate answer can also result from a task so difficult that the respondent simply gives up.

A request for clarification is usually a reliable sign that a question has problems. A good question should stand on its own without any need for explanation or additional information.

From these few examples we can see that following the trail from the behavior to its cause is often not easy. So while a high incidence of deviation from ideal interaction may suggest something is wrong, the method is not good at indicating exactly what is causing the deviation.

Beyond the basic codes in Exhibit 6.1, auxiliary codes are sometimes used. We might want to know, for example, how often interviewers have to probe to obtain an adequate answer, or whether they follow training guidelines when they handle respondent questions, or perhaps how often respondents change their answer to a question. Although there may be good reasons for using additional codes, we recommend that the novice survey researcher resist the temptation to add codes.

But, if there is a compelling reason to use additional codes, what criteria should they meet? First, they should capture important information. The researcher should have clearly in mind how the additional data will be used. If it is just something interesting to know, then, as with adding questions to a questionnaire, adding codes should be avoided. Second, they should be well-defined; that is, they should not overlap with any other codes. Third, auxiliary codes should be clearly enough described that coders will be able to reliably use them. Finally, although one can't tell this in advance, it should seem plausible that the auxiliary behavior may occur often enough to be of concern.

When too many codes are added, it increases the likelihood that the 15% threshold will be passed. So it is possible that the number of questions flagged as problematic is, to some extent, a function of the number of codes employed.

The analysis of behavior codes is often limited to a simple tally of the total number of deviations for each question. Under some circumstances it is possible to do something more. For example, it can be useful to examine frequency of deviation codes by interviewer. An outlier with a large number of deviations can affect the results. This is easily uncovered. If the pretest is large, other types of analyses are also possible. Do deviations that may be caused by interviewer performance, rather than the instrument, decrease over time or not? If misreadings decrease after many interviews, it may be that interviewer training and practice are more at issue than is question wording.

The main strengths of behavior coding are that it is inexpensive, conceptually simple, and quantitative. Coders can be trained in the basic scheme fairly quickly. Behavior coding can be easily used in conjunction with other pretest methods. Supervisors monitoring interviews in a conventional pretest can behavior code those interviews. As Fowler (1995) has pointed out, this can pay an extra dividend when used in conjunction with interviewers' subjective assessments of respondent problems. If in a conventional pretest debriefing the interviewers claim that many respondents had difficulty with a particular question, we would expect some evidence of that to show up in the behavior code data.

A shortcoming of the method is that it does not provide information about *why* the question is problematic. So, while it is an efficient method to flag potentially problematic questions, it provides too little information for improving the question. For this reason, behavior coding alone is generally insufficient for pretesting.

A Note on Intercoder Reliability

To what extent will coders be consistent in using behavior codes? Ideally, if a pair of coders listen to the same interviewer–respondent interaction, they should code it identically. One obvious measure is the percentage of times that the two coders agree. The shortcoming of percentage agreement is that there is some possibility that coders will agree by chance. A better measure of coder agreement is Cohen's kappa, a statistic that corrects for such chance agreement. A detailed discussion of Cohen's kappa is beyond the scope of this book. Readers who want a better understanding of this measure should see Cohen (1960) and Everitt and Hay (1992). When resources permit, it is useful after coders have been trained to have them all code the same audiotaped interview to assess intercoder reliability before production coding begins.

Cognitive Interviews

Still another approach to pretesting to determine respondents' comprehension is to do one-on-one or cognitive interviews. In this participating arrangement, respondents are recruited and asked to come to a central location. The questionnaire is administered to each respondent separately. Borrowing a procedure called *think aloud* from cognitive psychology, respondents are instructed to think out loud as they answer each question. The idea is to determine what things respondents consider in answering the question. This method can provide insights into comprehension problems as do other pretest methods, but in a more direct fashion because respondents are explicitly reporting what they think about while answering a question. In addition to comprehension, this method also has the potential to identify problems in other phases of the response process, for example, performance tasks, such as recall, or using the response options.

The instruction given to respondents is of this form (closely following Ericsson & Simon, 1993):

```
We are interested in what you think about as you answer some
questions I am going to ask you. In order to do this, I am
going to ask you to THINK ALOUD. What I mean by think aloud
is that I want you to tell me everything you are thinking
from the time you first hear the question until you give an
answer. I don't want you to plan out what you are saying or
try to explain your answers to me. Just act as if you are
alone in the room speaking to yourself. It is most important
that you keep talking. If you are silent for any long period
of time, I will ask you to speak.
```

The interviewer will occasionally have to remind the respondent to keep talking aloud using probes such as, "What were you thinking about?" or "How did you come up with your answer?" Typically, in the course of the interview, the interviewer also asks a set of prepared, that is scripted, probes asked immediately after an answer is given. Although this latter practice is at variance with the procedures used by cognitive psychologists to elicit verbal reports, it is widely used by survey researchers. The resulting method is a hybrid of the think aloud method and the Belson-type "intensive interview" (Belson, 1981).

The think aloud procedure is often supplemented with question-specific *probes*. In fact, some practitioners rely more on such probes than on think alouds (more on that later). Probes are questions that ask respondents about particular aspects of a question. For example, a probe might ask, "What do you think the term 'racial profiling' means in this question?" Or, more broadly, "Would you tell me in your own words what this question means." Probes can be either scripted, that is prepared prior to the interview, or spontaneously developed by the interviewer during the cognitive interview.

Conrad and Blair (2004) found that relatively inexperienced interviewers can be taught to use think aloud methods effectively, whereas techniques that rely more heavily on unscripted probing require much more experience to conduct well. Unless you have substantial previous experience as a cognitive interviewer, it is best to stick to scripted probes. In constructing probes "on the fly" a less-experienced interviewer can easily lead the respondent ("identifying" problems that aren't really there), or simply produce probes that are not very effective. One of the few available sources for training cognitive interviewers is *Cognitive Interviewing and Questionnaire Design: A Training Manual* by Willis (1994).

In learning how respondents are formulating their responses to a question, we can gain insight into how they understand it. For example, in the question, "How often in the last year did you drink alcohol? Would you say more than once a week, about once a week, once or twice a month, or less than once a month," a respondent, thinking aloud, might mention something like: "Last year I didn't drink very much. There was a party on New Year's Eve, then probably the last time before that was Thanksgiving break, when a bunch of us got together to watch the Notre Dame game. . . ." In examining this think aloud transcript (called a *protocol*), the researcher realizes that the respondent is apparently thinking about the last calendar year, rather than the 12 months previous to the interview. If the latter is intended, the question might be revised to "How often in the *last twelve months*, since [GIVE DATE ONE YEAR PRIOR TO DATE OF INTERVIEW], did you drink any alcohol?" An alternative approach is to ask directly about the respondent's understanding of the reference period with an explicit probe like, "When you answered the question, did you think about the last 12 months or the last calendar year?" Such a probe might be effective. It is also possible that, if the think aloud did not suggest a problem with the reference period, such a probe could waste time. More importantly, probes need to be formulated very carefully so that they do not suggest problems to the respondent that don't exist. Just as interviewer probes in regular interviews can have unintended effects, so might they have biasing effects in cognitive interviews. For this reason, it is best for the researcher who is new to cognitive interviewing to avoid constructing probes spontaneously during the interview and to stick with scripted probes that can be carefully examined beforehand for potential bias.

Cognitive interviews can be especially useful when the respondents' tasks or the question concepts are potentially difficult. The most complete discussion of these methods and their theoretical basis is given by Ericsson and Simon (1993). Another source of survey application examples is Royston (1989). There is a rapidly growing literature on the use of cognitive methods both for testing survey questions and for studying the response process (see, e.g., Schwarz & Sudman, 1996; Schwarz & Wellens, 1997; Sudman, Bradburn, & Schwarz, 1996; Tourangeau, Rips, & Rasinski, 2000; Conrad & Blair, 2004).

Another advantage of cognitive interviews is that many problems can be uncovered very quickly. However, because of the very small samples typically used in cognitive interview pretests, it is hard to know how frequently respondents are likely to have a particular difficulty. Of course, larger pretest samples can be used if time and resources permit, but there is little research on optimal sample sizes. More importantly, there is little in the literature on methods for combining the results of large numbers of cognitive interviews for quantitative analysis.

In cognitive interviews what the respondent was thinking at the point of answering the question is fresher in mind than it would be in the postinterview method. The disadvantage is that the resources required for this type of interview are greater. More people have to be contacted to find respondents willing to come in for the interview. Typically, in one-on-one interviews, unlike cold-call interviews, respondents are paid to participate. The remuneration generally varies from $10 to $50, depending on the amount of time required. Finally, cognitive interviews are usually conducted by relatively senior staff members whose availability is limited and whose time is more expensive. So our trade-off, given a fixed budget, is to do fewer of these interviews and more of the cold-call interviews.

The results of cognitive interviews are often analyzed very informally (Tourangeau et al., 2000). Frequently, the interviewer goes over notes she made during the interview, perhaps in concert with an observer who also made notes. Based on these observations they decide what questions are problematic, usually something about the nature of the problem, and sometimes consider ideas about how the question could be improved. It is very easy for such procedures to go beyond the interview data and to be based more on interviewer judgment about the question than on anything suggested by actual respondent comments or behaviors.

Conrad and Blair (1996, 2004) found that when independent analysts tried to determine from a cognitive interview transcript whether or not particular questions caused problems for respondents, the results were not very reliable. That is, if two (or more) analysts review the same transcript they frequently came to different conclusions about the presence or absence or problems. Despite these caveats and shortcomings, various versions of the cognitive interview method have been widely used, and deemed useful, for nearly 20 years.

We have focused on how cognitive interviews can uncover comprehension problems. However, it is possible for a respondent to have perfect comprehension, but have great difficulty doing the other tasks required to answer. For example, respondents might have little difficulty arriving at an acceptable interpretation of the question, "How many movies did you go to last year?" If a respondent goes to a lot of movies, the recall and counting task might be quite difficult and error prone. While conventional pretesting, for example, might well reveal this difficulty, cognitive interviews using think aloud are probably better for this. Moreover, in hearing respondents thinking aloud, some details, beyond the simple fact that the task is hard for some people, may emerge.

In a program of research led by Seymour Sudman, cognitive interviewing that relied heavily on think alouds was shown to be very effective in learning about how respondents answer behavioral frequency questions both for self-reports and proxy reports (see, e.g., Blair, Menon, & Bickart, 1991).

While cognitive interviewing has most often been used for interviewer-administered surveys, it is also possible to use variations on the cognitive interview technique for testing self-administered instruments, whether conventional mail instruments or Web questionnaires (Schechter & Blair, 2001). Much of the use of cognitive methods on self-administered instruments has been for organizational surveys (Willimack, Lyberg, Martin, Japec, & Whitridge, 2004). In addition, a program of research to examine new versions of the race and ethnicity questions for the U.S. census, large numbers of cognitive interviews were conducted in a series of studies (see Davis, Blair, Gourdreau, Boone, Johnson, & Robles, 1998).

The possibility of the cognitive interview affecting response performance is a concern for any administration mode, but especially for self-administered instruments. In responding to a written questionnaire, the respondent has to both read the instrument (typically to himself) and talk aloud. Most self-administered instruments also include instructions to the respondent, definitions, and skip patterns, which must be attended to. These multiple demands on the respondent might degrade performance and cause mistakes and difficulties that might not otherwise occur.

An important part of the cognitive interview process is to minimize its "interruptive" effects. For self-administered questionnaires, this can be done by relying less on concurrent and more on retrospective think alouds, in which respondents report their thoughts immediately after answering a question rather than while answering it. Probes should be saved for natural breaks in the instrument—the bottom of pages, the ends of topic sections, and the like.

In summary, our recommendation to the novice survey researcher is to use cognitive interviewing in the early stages of instrument development, to rely primarily on think aloud, and to use scripted probing judiciously.

Respondent Debriefing

Respondent debriefing is a slight variation on the postinterview interview. In this instance, respondents, after completing an interview, are taken back through the entire interview, question by question, and probed about each item. These debriefings are usually done face to face, which usually means bringing respondents to a central office. The initial interview is often tape-recorded (or videotaped). The tape can be played back as a way to help respondents remember what they were thinking at the time they actually answered the questions. The costs are very similar to those for the cognitive interview procedure, with the exception that lower-level staff can often be used, and the small addition of the cost of the recording equipment and supplies.

We should note that all these special methods require special training of interviewers beyond that needed to conduct the actual survey interview. Furthermore, while the methods are increasingly used in combination, although at additional costs, there is little research on optimal combinations. For studies with many new or potentially difficult items, we suggest beginning with a method that probes problems related to respondent comprehension and recall fairly directly, such as the postinterview methods or cognitive interviews, then moving on to conventional pretesting with behavior coding. For very simple instruments, conventional pretesting alone may be sufficient.

Expert Panel

Finally, Presser and Blair (1994) found that the use of an expert panel was very effective, compared to other methods, in identifying problems with a questionnaire. An expert panel is a small group of specialists brought together to critique a questionnaire. The participants are primarily survey professionals, but subject-matter experts may also be included. The group, consisting of three to eight members, is given a copy of the draft instrument several days prior to the meeting. Then, in a group session, the panel reviews the questionnaire item by item, a process similar to conventional interviewer pretest debriefing. The difference is that the panel's recommendations stem from their knowledge of questionnaires (or subject-matter issues) rather than from the reactions of pretest respondents. In an experimental comparison to other methods, Presser and Blair (1994) found that such panels were efficient in identifying many of the types of problems identified by other pretest methods, and also uncovered other questionnaire flaws (such as potential analysis difficulties). The strengths of expert panels come from their diversity of expertise as well as their interaction in the meeting. The Presser and Blair panels consisted of a cognitive psychologist, a questionnaire-design specialist, and a general survey methodologist. Clearly, other combinations also should be tried. It is important to note that consensus on either potential problems or solutions is sometimes not reached by the panel, but this is also true of other methods. However, we strongly suggest resisting the temptation to use experts as a substitute for pretesting with respondents.

Even the novice survey researcher, without substantial financial resources to pay panel participants, has options that permit the use of the expert panel. Because the amount of time required is small—usually only 3 to 4 hours in total—it is often possible to find people who will participate (once or twice) without payment. Faculty members at a university may be called on (particularly for a class project on campus), as may colleagues at other universities. Another resource may be survey professionals at nearby survey organizations. Finally, researchers who have done surveys in the subject area may be asked to help. (These last may be identified during the project's literature-review phase.) If travel time and costs are a limitation, it is possible to conduct the panel discussion relatively cheaply via a conference call, but a face-to-face gathering is preferable.

Assessing Interviewer Tasks

An interviewer's performance is also a crucial part of the survey's quality. As we have noted, the interviewer can be a major contributor to total survey error. Fortunately, given limited resources, many of the procedures that uncover sources of respondent error are also effective identifiers of interviewer error and problems.

Interviewer debriefings are obviously an opportunity for interviewers to report difficulties such as awkward questions, logic errors with skip patterns, words that are difficult to pronounce, or questions that come across differently when read aloud than when seen in print. In fact, one has to take care during the debriefing session that interviewer problems are not dwelled on at the expense of reports of respondent difficulties.

Monitoring telephone pretest (as well as main) interviews is a standard procedure at most survey organizations. Telephone-monitoring equipment allows a supervisor or investigator to listen to interviews in progress without either the interviewer or the respondent being aware that they are being monitored. This is done for quality control during main data collection, to see that procedures for administering questions are being followed. It is also a good way to detect rough spots in an interview during the pretest. We urge researchers to monitor live interviews. Questions sometimes come across quite differently when heard than when read; phrases that may seem moderately awkward or complex on the page often are unbearable in the interview. For the novice researcher, monitoring is also an invaluable aid to learning how to write questions that sound natural.

Revising and Retesting: Deciding
Which Pretest Problems to Address

It is not always possible, or advisable, to address every problem identified in pretesting. First, we may decide that an identified "problem" is not really a problem or is too trivial to warrant the cost of a change and the additional testing then required. It is common—often to the chagrin of the interviewing staff—that issues raised in pretest debriefings are not incorporated into the next draft of the questionnaire. In other instances, we may feel that the problem will seldom occur and that addressing it may introduce other problems. For example, assume that in pretesting the question "Most of the time you drank alcohol, on average, how many drinks did you have at one time?" a single respondent reports that when she is out with her boyfriend she sips from his drinks rather than having her own. Consequently, she finds it difficult to provide an answer in terms of "how many drinks" she had. It would require an extensive revision to accommodate such a respondent and in the process, a revision would make the question needlessly complicated for the majority of respondents. In this case, it makes sense to leave the item unchanged.

Those situations aside, we suggest returning to the question utility guide in Exhibit 4.2 to help decide on changes. If, based on these criteria, the problem discovered in the pretest affects the question's usefulness, the question needs to be either revised or dropped. Sometimes, we find that a question is simply too difficult to ask, perhaps because the level of complexity of the research goal requires multiple items. If the question is central to our research, then the multiple items are sensible. But if the item is less central, we may decide that the questions that need to be added are not worth the resources. In other instances, the question may not be suited to the administration mode: It may be too long or have too many response categories for a telephone interview. In such situations, if a question cannot be effectively revised, it is wise to drop it.

Deciding How Much Testing Is Enough

The different kinds of testing we have described can obviously become quite elaborate, time-consuming, and costly. We should anticipate the amount of pretesting we will need and devise a plan based on the complexity of the instrument, the proportion of new items, and, to some extent, the skip-pattern structure. If there are numerous branches to sections asked of only some respondents, larger pretests may be necessary to ensure complete testing of all sections of the questionnaire. After we have completed the amount of pretesting we decided we can afford and have evaluated the results using the question utility guide, we must make the hard choice either to drop items that do not work or to shift resources (both time and money) from other parts of the project to do further testing. This last course should be taken only if two conditions hold: the problematic instrument items are central to our research questions, and we will not affect other parts of the design (such as main sample size) to the point of questionable usefulness.

Pilot Tests

One last method of design and instrument development is the pilot test. In a pilot test, a relatively large number of interviews are conducted using the exact procedures planned for the study. Pretests are a necessary preliminary to pilot tests. As we have seen, in the pretest, we use a small number of cases to detect flaws in the questionnaire. We try to develop the instrument and interview procedures to maximize cooperation and to aid the respondent in the required response tasks. But there are aspects of the data collection process that pretesting alone may not address.

For example, pretest samples are usually too small to use to estimate cooperation or screening eligibility rates. Additionally, some field procedures may not be testable with small samples. There may be unknown characteristics of the sampling frame, or of the target population's response to aspects of the instrument, that can be examined only in a large field test. Problems that affect a small, but important, subgroup of respondents may not be detected with the small samples typically used for pretesting.

When the cost of the main study is high, or when some of the procedures are innovative or unusual, it may be risky to proceed to main data collection without a pilot study. The cost of pilot tests should be incurred, if at all, only after thorough pretesting. Making major changes after pilot testing is very inefficient.

Because most small-scale studies cannot afford pilot studies, it is best to stick to well-tested field and sampling procedures. To this end, it is useful to spend some time at the outset of the study searching the literature for both survey questions, as noted earlier, and for studies that have surveyed our target population. Consider how well the mode of data collection worked; whether there were cooperation or data-quality problems; and whether special interviewer-training procedures were used (e.g., in handling sensitive questions) that we might adopt.

Although there is little in theory or practice to inform us about how these methods might be optimally combined to maximize what we learn from pretesting for minimal cost, multiple methods are routinely used together.

Some Last Advice

The questionnaire is the indispensable means by which the opinions, behaviors, and attributes of respondents are converted into data. The focus in Chapters 4 to 6 is on the development and testing of the questionnaire, reflecting the instrument's central role in the survey process. Still, there is much more that could be said, and the first questionnaire the reader attempts to construct is likely to present issues that were not addressed here. Thus, it is very important to consult the references provided in this and other chapters. Often a point that was omitted or only touched on here is treated in detail elsewhere.

We end with four points. First, to the extent resources permit, use multiple pretesting methods and multiple rounds of testing. Every flaw that is found during testing is one less problem to confront during analysis and interpretation. Second, learn to assess survey questions by reading them aloud and developing your ear. A question that sounds awkward or unnatural or is hard for you, the researcher, to read will surely cause problems for respondents. Third, consider no question's wording sacrosanct. If a question does not pass muster in pretesting, its flaws must be confronted, however difficult that may prove or however enamored you may be of its original form. Finally, seek out examples of questionnaires written by experienced researchers. You cannot consult too many examples while learning the art of questionnaire design. In addition to texts and journal articles, copies of questionnaires from surveys conducted by major survey organizations or federal government statistical agencies are often available for free (even through online databases).

It has been remarked that all computer programs have one thing in common: They can be improved. The same is equally true of survey questionnaires.

Note

1. See Converse and Presser (1986) for a discussion of the role of vignettes in surveys, sometimes called factorial surveys. When this approach is used, the factors (in brackets in the example) are varied. With multiple factors, the number of versions can increase very rapidly. Careful pretesting is needed to decide which factors will serve to separate groups of respondents with different positions on the issue.

7

Designing the Sample

M ost of what we do in surveys relies on common sense. In sampling, for example, we need to think about which population we want to study, what list or resource we can use that includes this population, how good this resource is, what problems we might encounter, and how we can overcome them. Knowing a few basic principles and using common sense can greatly help the researcher in addressing these concerns.

The first part of this chapter defines sampling, explains why it is more efficient than interviewing everyone, and illustrates the differences between non-probability and probability samples. Because probability designs are the preferred method, the remainder of the chapter focuses on them. Particular attention is given to the key tasks of defining the population, constructing and evaluating sampling frames, and handling unexpected situations and common problems. We also address the commonly asked question, "What sample size do I need?" Finally, a number of examples are presented showing how to use census data to plan a survey and to determine whether the number of interviews with important subgroups will be adequate for analysis.

The Basics

Sampling is the selection of elements, following prescribed rules, from a defined population. These elements are usually the subjects of the study: They can be individuals, households, farm animals, sections of land, business transactions, hospitals, and so on. There are two main reasons for sampling. One is to generalize to or make inferences about the population of interest for research questions (e.g., How many people were unemployed but looking for work in March? What proportion of adults in Maryland believe that violent crime in their neighborhood has gotten worse?). A well-executed probability sample allows us to estimate these numbers and percentages.

A second reason for sampling is that it is more efficient and less expensive than a **census,** which attempts to include or ask questions about every element or member in the population. It took about 10 years to plan, carry out, and report the results of the *1990 Census of Population and Housing* at a cost of about $2.6 billion. The 2000 Census is estimated to cost in excess of $5 billion. From 1988 through 1990, the Census Bureau created about 635,000 temporary jobs and hired about 565,000 people (U.S. Census Bureau, 1989). For 2000, one report estimated that the Census Bureau would need more than double the number of enumerators used in 1990. Think about the difficulties involved in hiring this many people, especially well-qualified people. Then think about training them and building in quality-control checks. Although the Census Bureau does an outstanding job in accomplishing this feat, the logistics of this task contribute to proportionately more errors, higher costs, and a less-capable staff than would a large sample survey. In fact, this alternative has been discussed on a number of occasions, but because it would require a change in the U.S. Constitution, it is very difficult to convince members of Congress and state legislators of the advantages of a sample survey.[1] So, the debate goes on.

There are two types of samples: probability and nonprobability samples. **Probability samples,** or *random* samples, are those in which every element has a known, nonzero chance of selection and the elements are selected through a random procedure. Although elements do not need to have an *equal* chance of being selected, every element must have *some* chance, and that chance must be *known.* By fulfilling these two conditions and using the correct statistical formula, we can estimate values for the entire population and the margin of error of that estimate. For example, assume we want to estimate the number of students on a university campus who had at least one cold during the past winter. (The potential measurement problems inherent in this seemingly simple research question are ignored here but are discussed in Chapter 4.) Let's assume that winter is defined for our purposes as November 15 through March 15; that the school has a total of 28,000 enrolled students; that we can obtain a list of all students; that this list has no missing or incomplete information; that we want to select a sample of 1,000 students; and, finally, that all students who are sent a questionnaire will respond (perfect cooperation). A random sample of 1,000 students out of 28,000 gives each student a 1 in 28 chance of being in the study or a **probability of selection** of .0357. This is just saying the same thing in two different ways.

Probability of selection = 1,000/28,000 = 1/28 = .0357

If 62% (620) of the students in the sample say they had a cold last winter, we can estimate the percentage or the number of students in the total population who had a cold last winter by using the probability of selection to compute an estimate of the range of possible values wherein the true population value may lie, otherwise known as the **sampling error.**[2] To make this computation, we multiply the reciprocal of the probability of selection (1/.0357) by the number of students who said they had a cold, and we compute a **confidence interval** for this number. A confidence interval for a numeric variable is a range of values above and below the sample estimate that should contain the population value. To compute the confidence interval, the researcher sets a probability—for example, 90%, 95%—that the confidence interval includes the true population value. The reciprocal of .0357 is 28. Does this look familiar? We multiply this number by the number of students who had a cold (620). This gives us an estimated total of 17,360 (620 * 28 = 17,360). Next, we must compute a confidence interval because this is only one sample out of many, many possible samples of 1,000 students that could be selected from a population of 28,000. Using a 95% probability level, the confidence interval for our sample estimate is 18,202 to 16,518. This means that if we were to survey repeated samples of 1,000 students as in our example, the confidence intervals for 95% of the samples would contain the true population value for the number of students who had a cold last winter. We will say more about estimating population values later. The major point to understand from this exercise is that to have a statistical basis on which to estimate a population value from a sample, the sample must be a probability sample.

A second type of sample is a nonprobability design, which includes convenience, purposive, quota, and snowball samples. The names for these samples suggest the method by which respondents are selected for the study, as the following examples illustrate.

In a study to see who studies more outside of class—juniors majoring in engineering or juniors majoring in a social science—a convenience or purposive sample could be used to select two junior level classes in engineering and two junior level classes in the social sciences without a random procedure or known probability of selection. We would ask all the students in each class how many hours of homework they do in an average week. The classes would be purposefully selected—junior level classes in engineering and social sciences—and we might select classes that would be convenient for our schedule or because we know the instructors and it would be easy to gain their cooperation. In these situations, all juniors in both disciplines would not have a known chance of selection, and there is no way of estimating how representative the results are of all junior engineering or social science majors.

In a quota sample, before interviewing begins we determine what characteristics of the population being studied, such as gender, race, age, and employment, might be related to our dependent variable. Before we conduct the study, we want to know what proportions of the population have these characteristics so that our final interviews reflect these population distributions. If the population is 50% male and 50% female, we want half the final interviews to be with males and half to be with females. If the study is being conducted in a specific city or county, we may find information about the population from U.S. census data or a recently conducted study. For studies of students, information is usually available from the registrar or the office of records and admissions. Let's assume that we are doing the study in a city and that gender and race are important variables. Census data indicate that for adults 18 years of age and older, 52% are females, 48% are males, 27% are African American, 68% are white, and 5% are some other race. Assume that we need 200 face-to-face interviews and that the distribution of the population by race is the same for both genders. Knowing these percentages, we can now estimate interview quotas for the sample:

	Number of Interviews	
	Women	Men
White	71 (.68 * 104)	65 (.68 * 96)
African American	28 (.27 * 104)	26 (.27 * 96)
Other	5 (.05 * 104)	5 (.05 * 96)
Total	104 (.52 * 200)	96 (.48 * 200)

Interviewers would be assigned the above quotas, and they would be sent to designated areas, such as street corners, shopping malls, or selected city blocks. At the assigned areas, the interviewers would be instructed to fill their quotas. When the study is completed, the interviewed sample usually matches the population exactly on the important variables.

There are three major problems with this design. One is the lack of control over who is interviewed. The interviewer is told the *type* of person to interview, but not who. For example, at a shopping mall, interviewers can exclude specific individuals whose looks they do not like or who seem to be in a bad mood or in a hurry. Thus, those who are included may be different in important ways from those who are excluded. A second problem is that records are usually not kept of the number of people contacted or the number who refused to be interviewed. The only records are the number of completed interviews, and we know that the sample characteristics of gender and race match the population. Again, the group that agreed to be interviewed may be different from the rest of the population. They may be more cooperative, they may have more money, or, depending upon the day and time that interviews are conducted, these respondents may overrepresent the elderly or unemployed people. We also have no assurance that other important variables, which were not used in defining the target groups, will be distributed in our sample as they are in the population.

The third and most critical issue is that the probability of selection of these 200 respondents is unknown. If the study was done at a shopping mall, those in the eligible population who never visit that mall would have no chance of selection. Conversely, people who visit the mall two to three times a week would have a greater chance of being in the study than those who visit it less often. These same criticisms apply to "street corner" and other types of haphazard designs. In short, if you wish to generalize to a population, all the elements in the population must have a known, nonzero chance of being in the study.

Another frequently used nonprobability design is snowball sampling, used when the sample units are rare or hard to find. One assumption underlying this method is that people with similar characteristics or attributes are likely to know each other. A more general assumption is that individuals without the characteristic may know others who have it. Suppose we are interested in interviewing people who are HIV-positive to see what aspects of their lifestyle they may have changed to forestall the onset of AIDS. Clearly, this is a rare and hard-to-find population group. No sampling frame exists for this group and, although its numbers are increasing, it still represents a very small percentage of the total population. To conduct such a study, we might combine two nonprobability designs. First, we need to locate or gain access to organizations or groups where HIV-positive people can be found—for example, organizations of gay men or AIDS support groups. Depending on the survey's time schedule and resources, a convenience or purposive sample of organizations might be contacted. Access to some or all of an organization's members is obtained usually by posting announcements on bulletin boards or by giving a presentation to members at an open meeting. A few respondents will volunteer to these requests, and this gives the researcher a starting point. The next step is to conduct the interviews. At the end of the interview, using the "snowball" tactic, the interviewer asks the respondent for names and telephone numbers or addresses of people the respondent knows who satisfy the eligibility conditions for the interview; that is, identified members of a population are asked to identify other eligible people.

While there is truth to the underlying premise of the snowball design—that people with similar characteristics associate with each other—it also has the flaws typical of a nonprobability sample. How well does the final sample represent the population of HIV-positive people? We are unable to answer this question because we don't know the probability of selection and can't estimate the boundaries or limits of sampling error. In addition, we don't know how limiting our initial selection of organizations and volunteers was. Do the cooperating organizations or volunteers know all the other HIV-positive people in the study area or are organizations and friendships closed cliques? The inability to estimate a range of error for our sample results seriously undermines the credibility of the final study.

Nevertheless, nonprobability samples can serve useful purposes. Not every research study is designed to estimate some characteristic of or generalize to a population. In an exploratory study, a researcher may only want to get a sense of what respondents are thinking, believe, or feel about a topic, information that may be useful in designing a larger and more comprehensive study at a later time. Nonprobability samples are also useful in the early stages of testing questionnaire items. We may be uncertain how specific types of individuals will interpret the wording of certain questions, and thus, we may want to focus our efforts only on these individuals. Or, if we have prior knowledge about the population or groups of interest, we can use that information to better focus the study resources. For example, if we are developing an education module for health care workers dealing with the most misunderstood aspects of AIDS, we might survey groups that have some knowledge about AIDS rather than to survey all health care workers. The assumption in this case is that misunderstandings of a group such as nurses who have cared for AIDS victims will also be present in other groups of health care workers. Thus, the results of a study of AIDS nurses would be generalizable to all health care workers.

Defining the Population

Let's return to a discussion of probability samples. Two key tasks in creating probability samples are defining the population and selecting or constructing the sampling frame. The population is the group or aggregation of elements that we wish to study, the group to which we want to generalize the results of our study. We begin the process by referring to the research problem. What is it that we wish to study? We may want to know, for example,

"What proportion of adults believe a woman should be allowed to have an abortion for any reason?"

There are two components of the population definition that must be specified: the units or elements of analysis and the defining boundaries of the units. For units of analysis we must decide if the research problem focuses on individuals, households, group quarters, or something else. The term *individuals* is self-explanatory, but the other two terms may not be. According to the U.S. Census Bureau (1993) definition:

> A household includes all the persons who occupy a housing unit. A housing unit is a house, an apartment, a mobile home, a group of rooms, or a single room that is occupied as separate living quarters. Separate living quarters are those in which the occupants live and eat separately from any other persons in the building and have direct access from the outside of the building through a common hall. The occupants may be a single family, one person living alone, two or more families living together, or any other group of related or unrelated persons who share living arrangements. (p. B-13)

Group quarters are all other types of living arrangements. The Census Bureau has defined two types of group quarters: (1) institutional and (2) noninstitutional. Institutions are places where occupants are under "formally authorized, supervised care or custody" such as prisons, federal detention centers, military stockades, local jails, police lockups, halfway houses, nursing homes, mental hospitals, schools or wards for the mentally retarded, homes for neglected children, and training schools for juvenile delinquents. The Census Bureau defines noninstitutional group quarters as living arrangements of ten or more unrelated people, such as rooming houses; group homes that provide care and supportive services, such as maternity homes for unwed mothers; religious group quarters; college dormitories; agricultural workers' dormitories; emergency shelters for homeless people; crews of maritime vessels; and staff residences of institutions. Investigators must decide in the design phase of the research which types of residential living arrangements contain the eligible respondents for their research.

How do we decide whether we want to study individuals, households, or some other units? It depends on the objectives of the research problem. If we are studying attitudes about abortion, individuals would be the appropriate units of analysis. Because attitudes might differ by gender or age, it would not make sense to designate households as the units of analysis. If attitudes differed within households, for example, between spouses or between adults and children, there would be no convenient way to report the attitude of the household. While attitudes are individual in nature, total income or the number of durable goods owned or purchased in the last year, such as refrigerators and automobiles, may best be studied by household. Although individuals and households are common units of analysis, other possibilities include organizations or institutions such as labor unions, banks, colleges, environmental groups, hospitals, and clubs. The appropriate unit of analysis for studying the types and quality of science facilities in elementary schools is elementary schools. Hospitals would be the unit of analysis for estimating the number and types of organ transplants within the last 5 years. Deciding the unit of analysis is up to the researcher, and it is a decision that must be made in the design phase of the research.

The second task in defining the population is to specify the defining boundaries. For a population of individuals, this usually involves specifying the geographic boundaries of the study area, the demographic characteristics of the study population, the age of respondents, and the residential units included. The geographic boundaries of the study area are determined by the population the study will generalize to and the amount of available time and resources. Depending upon these constraints, the study area may be defined as a city, county, metropolitan statistical area, state, or some other geographic entity. The geographic area must be specified early in the design phase because it has implications for many other activities, such as obtaining a sampling frame, hiring interviewers for a face-to-face survey, and estimating telephone charges for a telephone study.

We must also define the characteristics of eligible respondents and the residential units in which they may live. For example, the General Social Survey (GSS) is a national probability sample of English-speaking adults who reside in households in the 50 U.S. states. There are a number of points to note about this definition. First, this survey includes all 50 states. For many face-to-face studies, Alaska and Hawaii are excluded because of the higher costs to train, supervise, and maintain interviewers in these states. These two states comprise only 0.065% (less than 1%) of the U.S. population (U.S. Census Bureau, 2000d), and many investigators are willing to live with the more restricted definition of the population. Second, the survey includes only English-speaking adults. This excludes approximately 2.5% of the sample households for the years 1987–1991 (Davis & Smith, 1993). The excluded households represent about 13 language groups, 60% of which speak Spanish. To include these groups in the survey would require a number of activities that would significantly raise the cost per interview. To include non–English-speaking respondents, we first need to anticipate what language groups or dialects may fall into the sample. Will they be German, Chinese, Turkish, Arabic, Vietnamese, or some other language group? After anticipating the groups, we need to have a double translation of the questionnaire performed for each language. The questionnaire is translated from English into the foreign language and then the foreign language version is translated back into English, ensuring that language subtleties and innuendoes have not been overlooked. Additionally, interviewer-training manuals need to be developed and/or modified and bilingual people in each of the languages must be hired, trained, and sent to the non–English-speaking households. Clearly, these activities not only increase the survey costs but may also increase the time required to complete the work.

A third consideration is the age of an eligible respondent. Adults may be defined as any person age 18 years or older, 19 years or older, or 21 years or older. It all depends on the objectives of the research and the group to which we wish to generalize. Sometimes only people within a specified age group are eligible, such as all persons age 18 to 64 years or women age 21 to 49 years. In the former age group, the interest may be in adults who are not retired; in the latter, the goal may be to study women after high school to see whether they joined the labor force, at what age, and in what occupations.

The type of dwelling unit in which the eligible respondents reside is a fourth consideration when defining the population. The GSS surveys households; this means that people residing in dormitories, shelters, nursing homes, and in other types of group quarters are excluded from the survey. In 2000, 2.76% of the U.S. population resided in group quarters, but the percentage varies considerably by age group (U.S. Census Bureau, 2000d). In 1980, for example, people living in group quarters represented approximately 9.4% of people age 18 to 24 years and 11.4% of those age 75 years and older, but only slightly more than 1% of the population between the ages of 25 and 64 years (Davis & Smith, 1993). In 2000, 0.4% of those younger than age 18 years lived in group quarters, 3.1% of those age 18 to 64 years, and 5.7% of those age 65 years and older.

Decisions about how broadly to define the population depend on several considerations. A number of questions must be asked: Which groups do I want to generalize to? How have other researchers defined the population when they have investigated this issue? What are our resources for doing this study, not only in terms of time and budget but also in personnel and facilities? Is there a "best method" of data collection for the research issue or the defined population? Is there a list or resource available that contains the eligible members of the defined population?

Constructing a Sampling Frame

Once we have defined the population in probability sampling, the next task is to find or construct a sampling frame. The frame is the list(s) or resource(s) that contains the elements of the defined population. For a telephone survey, the frame may be a telephone directory or a list of all the telephone prefixes for the geographic area. For mail surveys, the frame is usually a list with the names and addresses of all the elements, such as an organization's membership list. If the research objective is to determine the satisfaction of recent graduates with the academic training they received at a college or university, a list of the names and addresses of graduates in the past 5 years might be obtained from the alumni office or from the registrar's office. Can you think of any problems with using these types of lists? (We discuss these issues shortly.)

A frequently used sampling frame is a compilation of census data. These data are used when no reasonably complete list(s) exists for the defined population. Census data are usually used as a frame for face-to-face, general population studies, for example, of adults 18 years of age and older. You might believe that lists of all adults in the United States do exist. However, there are only two relatively complete lists, and both are confidential and not available to the public: census forms that are completed every 10 years, and income tax forms that are filed each year. Other lists that might be available all have major exclusion problems. For example, telephone directories do not include households without telephones or those with unlisted telephone numbers; lists from state driver's license bureaus do not include people who do not drive; voter registration lists exclude about one quarter to one third of all adults; and property tax records or utility company files are not inclusive of all adults and may not be public information.

How do you select a sample of adults using census data when the names and addresses are confidential? That is a good question. A brief example that avoids complex sampling theory will be useful. (Because parts of the process described involves sampling theory that is beyond the scope of this book, you will need to accept certain decisions on faith.)

We begin with the research problem. Assume that we want to study the attitudes of adults toward abortion. What is the next step? We need to define what we mean by *adult* and specify the defining boundaries. Let's assume we are interested in interviewing people age 18 years and older who reside in households and that we have enough resources to conduct a statewide face-to-face interview survey. Assume further that our required sample size is 1,000 completed interviews and that the average number of completed interviews per city block or final stage of sampling is five.[3]

We select the sample in stages, using census information. At the first stage, we select counties or groups of counties; for this project, we want to sample 24 counties. The 24 counties will be selected with probabilities proportionate to the number of people 18 years of age and older in each county.[4] Within each county, we will select city blocks or sections of land in rural areas. This stage will also be selected with probabilities proportionate to the number of adults.

Note that we have conducted two stages of selection based only on census data, and without names and addresses. At the final stage, however, we must count the households in the blocks or sections of land selected in the second stage. A person called a *lister* is sent to each block or area to compile a list of the addresses or locations of every household. A comparison is then made between the number of households found by the lister and the number that was used to select the block or area (census data). If we expect a 70% response rate, then approximately seven households per block ($5/.70 = 7.14$) are selected for interviewing. Using counts of people or households within defined geographic areas, and then at the last stage of selection going out and making a count, is one way in which a probability sample can be selected without having a list of the population before you start. One final note of caution: The numbers used in each stage of selection must be reasonably accurate or unknown biases can result.

Matching Defined Populations and Sampling Frames

Many times a perfect match is not possible between the defined population and an available sampling frame, and compromises must be made. A telephone survey of adults illustrates this problem. Because every adult does not have a telephone and many have unlisted numbers, a perfect frame does not exist. Let's go back to our study about the attitudes of adults toward abortion to illustrate the point.

Let's assume that we have decided to do the study in Wake County, North Carolina, the county that includes the city of Raleigh, and we have decided to do the study by telephone. What can we use as a sampling frame that would give every adult in the county a known chance of being in the sample? The first thought, and actually a good one, is the white pages of the telephone directory. However, as a frame, the telephone directory has a number of weaknesses. First, it excludes people without telephones. However, when we decided to do this survey by telephone, we accepted this potential bias. In Wake County, North Carolina, 1.2% of the occupied households do not have a telephone. The key question is, "Are the individuals in these households different from those who do have telephones for the variables being investigated?" In general, the answer is yes, they are different. However, in this case, they are a small percentage of the total and the cost to include them (by face-to-face interviews) far outweighs their value to the final results. For these reasons, we will live with this potential bias.

Second, a telephone directory excludes people who do not list their telephone number in the telephone book. In the Raleigh-Durham-Cary Combined Statistical Area, which includes Wake and seven other counties, this group represents approximately 22% of all households. Clearly, this number is too large to ignore. Random-digit dialing (RDD) sampling procedures, however, which are described in Chapter 8, overcome this problem. A third problem with the telephone directory as a frame, or with telephone surveys in general, is that some households have more than one telephone number, and hence, higher probabilities of selection. This problem can be offset by adding a question to the survey that asks how many different telephone numbers the household has, excluding business numbers. The survey data from each household with more than one residential number are weighted by the reciprocal of the number of residential numbers so that when we analyze the data, each household has the same probability of being in the survey.[5]

A fourth problem involves the geographic coverage of the Raleigh telephone directory. How much of the county is included: Is all, most, half, or maybe even more than Wake County included? This needs to be determined. Inspection of the telephone book indicates that there are two residential sections. One section includes the city of Raleigh and a number of surrounding towns and unincorporated areas. This section covers approximately 68% of the county and is serviced by one telephone company. The second section covers an area that is serviced by at least two other companies. This area includes the remainder of the county and a few small cities (population less than 5,000) and rural areas outside the county.

We have three options. One is to use only the first section of the telephone book, which includes Raleigh and approximately 68% of the county's population. We could redefine our population as adults that live in the city of Raleigh and in nearby areas. At the conclusion of our study, we would need to describe the characteristics of this population. We might note the proportion of adults living in cities or urban areas, the proportion in rural areas, the median income of this population, the mean years of education completed, and so on.

A second option would be to use both sections of the telephone book. To make the sampling frame match our population definition, we would need to screen some of the telephone numbers. Although the telephone directory lists all the telephone prefixes in the county and notes the prefixes unique to each city, it does not indicate whether the prefixes extend beyond the city and town boundaries. Thus, for interviews in all towns that are on the county border, it would be necessary to ask each respondent whether his or her household is in Wake County or an adjacent county. From census data, we estimate that approximately 8% of the sample numbers would have to be screened in this manner. If the respondent's household is not in Wake County, it would be ineligible, and we would not attempt an interview.

A third option is to define our population frame as both sections of the telephone directory. Once again, after consulting census data, we estimate that approximately 4% of the telephone numbers in the book come from the adjacent counties of Franklin, Johnston, and Harnett. The simplest and most complete procedure would be to define the population as households listed in the Raleigh telephone directory—that is, all of Wake County and a few small towns and rural areas beyond the county border, which add an additional 4% to the population size. Because the county of Wake has no unique relationship to our research problem, there is no need for the sampling frame to exactly match the county boundaries. It is important, however, when we report our survey results, to provide the reader with a precise and accurate definition of the survey population and to describe the demographic characteristics of this population.

Recognizing Problems with Sampling Frames

One thing we have learned in our many years of survey experience is never to trust a sampling frame. Because incomplete and inaccurate information are inherent in sampling frames, the researcher needs to recognize the potential problems and to know how to handle them when they occur. The following problems often arise when the frame is a list:

- The list contains units that are not members of the defined population—these units are called *ineligibles*.
- Information about individuals or units on the list is inaccurate.
- Information about individuals or units on the list is missing.
- Some individuals or units are listed more than once.

Ineligibles

A common problem is ineligibles. What should you do about them? The answer is simple: Ignore them for purposes of interviewing, but include them in the sampling rate. The **sampling rate** is the number of selections that must be made from a frame to achieve the desired sample size divided by the total number of selections in the frame. An example helps to explain this. Assume we want to interview recent graduates from our university to determine their satisfaction with curriculum changes that were instituted 5 years ago. We want to interview only graduates of the past 3 years, but the only frame we can get includes people who have graduated in the past 6 years. What do we do? We use the 6-year frame, but we determine what proportion of the graduates come from the last 3 years and what proportion come from the previous 3 years. To make matters simple, let's assume that the proportions are equal, 50% from each period. If there is no order to the list by year of graduation—that is, people from both periods are randomly distributed throughout—we would expect a random sample of 1,000 names to include approximately 500 people from each period. Because we are not interested in interviewing people who graduated more than 3 years ago, every time such a person is selected, we simply ignore that person and go on to the next selection. However, we still need to estimate the number of ineligibles in the list so they can be accounted for in the sampling rate. Having ineligibles on the list and ignoring them does not affect the probability of selection. Here is why. If 18,000 students graduated in the last 6 years and 9,000 graduated in each 3-year period, the sampling rate would be 1,000/18,000 (1 in 18) or .055. We assume we will end up with 500 eligible graduates from the total of 9,000 eligible graduates for a probability of selection of 500/9,000 = .055. Ineligibles are part of the sampling rate, but they do not affect the probability of selection, at least not in this situation.

Frequently, however, the probability of selection *is* affected by ineligibles because of incorrect sampling procedures. Another example illustrates this. Assume that eligibles and ineligibles are mixed together on a list. Instead of adjusting the sampling rate based on the proportion of ineligibles on the list, researchers make the typical mistake of ignoring the ineligible person and selecting the next eligible person on the list. Can you guess what is wrong with this procedure? It gives eligible people who are preceded on the list by an ineligible person a double chance of selection: their own probability of selection and the ineligible person's chance of selection.

This incorrect procedure is commonly used when sampling from a telephone directory. Although the objective is a sample of telephone numbers, the sample selection process usually takes a sample of lines and telephone numbers. Because many telephone directories mix residential listings with business listings, to select a sample of households, you must treat business listings as ineligibles. Another common occurrence is that people with long names or addresses take up two lines. There are also other reasons why not every line in a telephone book has a telephone number (see Exhibit 8.2 in Chapter 8).

No one counts all the eligible telephone numbers in a telephone directory and then selects a simple or systematic random sample of the listings. Think about the telephone directory for a large city such as Chicago. It contains more than 1,400 pages of listings, with 5 columns per page. Every column comprises more than 110 lines. It might take a week to count all of the people in the book. What researchers do, instead, is to estimate the proportion of lines that do not contain an eligible household number, include this proportion in the sampling rate, and then select a sample of lines from the telephone directory. Although this is the correct procedure, you must be careful to give each listing the same chance of selection. So, if business and residential listings are mixed together, you need to estimate the proportion that are business listings. The same must be done with blank lines or other ineligible lines. In essence, an estimate of the total number of ineligible and blank lines must be obtained before a sampling rate is established. This estimate becomes part of the sampling rate.[6] The key is to ensure that the probabilities of selection for the elements do not change because of the selection procedures.

Inaccuracies

A second common problem with sampling frames is that they contain inaccuracies. Thus, we may select an element that we thought was eligible but it turns out to be ineligible. Because all sampled cases must be worked completely, and we do not know in advance which cases are eligible, ineligibles become a problem because each one takes time and money to work, reducing the number of eligible cases that can be obtained within the allotted time and budget. Therefore, it is important that this type of problem be found early, before or at pretesting, so that corrective actions can be taken.

Another potential problem with sampling frames is that, instead of finding one eligible element at a household or telephone number, more than one eligible element is found. Assume that we listed all households on a city block and then selected a sample of households at which to conduct interviews. The interviewer arrives at one address and, instead of finding one household, finds two: a front, listed apartment and a back, unlisted apartment. What should the interviewer do? In this case, the correct decision is to interview both households because the second household is eligible, even though it was not part of the initial frame. We want all eligible units to have a chance of being selected into the sample, and this is the only way the second unit has a chance of selection. In addition, we want to generalize to all eligible units, and this second unit is a member of the eligible units. In most situations, we can usually assume that mistakes are random. Because we happen to find one (random) error, there are likely to be others. It is also likely, as illustrated above, that some elements that were thought to be eligible are ineligible. If the frame is current and of good quality, these situations should be infrequent. However, knowing the proper corrective actions is still necessary.

Missing Information

Surprise elements can become a major problem. In a study designed by the authors in 1987–1988, the best available sampling frame was census data. We decided to use the census data but to update it in areas of high growth because it was already 7 years old. We used a number of sources: statistics compiled by the city on new housing starts and demolitions, and we contacted banks and real estate agents to ask them about areas of new growth. From past experience, we knew that no one source would have all the information we needed and that using multiple sources would be the best strategy. Using information from these sources, we updated the 1980 census data and made our block selections based on these data. Almost all the selections worked out well except for one block. We estimated that this block had about 50 households on it. When the lister arrived at the block, he found that a high-rise building had been erected and was now occupied. Instead of 50 households, there were now 450 households. In the previous example, when two households were found where one was expected, both were interviewed. Following a similar procedure in this case would mean using the same sampling rate for the 450 households that was to have been used with the 50 households—1 in 7 households. If we used a 1 in 7 rate, instead of selecting 7 households, we would end up selecting approximately 64 households. With a 70% cooperation rate this would yield about 45 interviews from this one block. Because we planned on doing a total of 500 interviews, this one block would contribute a disproportionate amount to our total number of completed interviews. In a situation like this, we might allow a double or triple number of interviews to be conducted in this block and then weight the results to reflect the number of interviews that should have been done in the block. If, for example, we allow 15 interviews to be conducted from that block, we would weight those interviews by a factor of 3 (45/15 = 3).

Other surprises can occur when sample respondents are selected. In most general population surveys of individuals, only one respondent per household is selected for interviewing. In a study of individuals, what do you think of this? Do all eligible respondents have the same probability of selection? The answer is no. In a one-person household, the one person is selected for interviewing. In households with three eligible people, only one person is selected, and the probability of selection for each person is one-third, at this stage. Thus, the probability of selection differs by the number of eligible people within the household. To make the probabilities equal for all individuals, the data must be weighted to reflect the number of eligible people in the selected households. In a one-person household, the weight is one; in a two-person household, it is two; in a three-person household, it is three, and so on.

Multiple Listings

Another common problem with frames is that some individuals are listed more than once, giving them higher probabilities of selection, unless some corrective action is taken. This situation occurs in telephone samples of households because some households, for a variety of reasons, have more than one telephone number. There can be one telephone number for the teenagers and one for the adults; the household may have a second telephone number that is given only to very close friends or relatives; or one of the numbers may be used for a business. The Federal Communications Commission estimates that approximately 6% to 8% of households in the United States have more than one residential telephone number.[7] The way to correct this situation is to add a question to the survey that asks about the number of different telephone numbers in the household and to weight the respondent's data by the reciprocal of the number of nonbusiness telephone numbers in the household. If there are two, the data are given a weight of .5; if there are three, it is .33; and so on.

A similar situation results from the use of university directories, which list the names and addresses of faculty, staff, and students. Some students or staff members may be listed in more than one section of the directory; graduate students who have assistantships may be listed as both students and staff. Many times staff members take courses, and thus they may be listed as both staff and students. In these situations, two things can be done. Every time a graduate student or staff member is selected, we would look in the other section of the directory to see if this person is listed a second time. We can assign all those who are listed twice a weight of .5, or we can randomly delete one-half of these cases. These procedures equalize the probabilities of selection.

Finally, one question that is frequently asked and for which there is no definitive answer is, "How accurate and complete should the sampling frame be?" The major problem with inaccuracies is that it takes time and money to work each case to find out that a respondent is ineligible or maybe that the respondent is eligible but has moved to a new location within the target geographic area. It usually takes additional time to obtain current information and then to attempt an interview. These efforts take resources that could be used to interview eligible cases. Thus, time and effort spent on inaccurate information usually reduces the number of eligible cases that can be interviewed.

How complete the sampling frame should be is also a difficult question. The researcher must ask a number of questions. How different are the missing elements from the included elements? How might the final results be affected because of the missing elements? Will only the outcomes for single variables be affected or will relationships between variables be affected?

Before adopting a frame, we need to investigate its inadequacies and assess how they may affect our results.

Determining Sample Size

One of the most frequently asked questions in survey research is, "What sample size do I need?" As usual, there is not a simple answer to this question. Sample size is a function of a number of things: the research design being used; the variability of the key variable(s), if we are trying to estimate a population value; or, if we are testing hypotheses, the size of the differences between two variables and the standard error of their difference. When testing hypotheses, the researcher tries to set a sample size that minimizes making two types of errors when drawing conclusions from the data: (a) to claim that variables are related when they are not (called Type I error); or (b) to conclude the opposite, that two variables are unrelated when, in fact, they are related (called Type II error). Resolving these issues and determining sample size is a complex problem that is discussed in detail by Cohen (1988), Fleiss (1981), and Moser and Kalton (1972). To give an idea of the information required to begin solving for sample size, we illustrate the classic method of determining sample size when estimating a population percentage.

An approximate formula (Cochran, 1977) for determining the sample size for a variable expressed as a percentage is

$$n = \left(1 - \frac{n}{N}\right) \times \frac{t^2(p \times q)}{d^2} = \frac{\textit{finite population}}{\textit{correction}} \times \frac{\textit{probability level} \times \textit{variance}}{\textit{confidence interval}}$$

Where

n = The sample size or the number of completed interviews with eligible elements

N = The size of the eligible population

t^2 = The squared value of the **standard deviation** score that refers to the area under a normal distribution of values

p = The percentage category for which we are computing the sample size

$q = 1 - p$

d^2 = The squared value of one-half the precision interval around the sample estimate

Expressed in words, the formula states that sample size is an expression of a finite population correction factor $(1 - n/N)$ times the probability level for this sample occurrence times the **variance** or the variability of our variable in the population divided by the size of the confidence interval that we want for our estimate. Let's talk more in depth about these components and then look at an example.

Finite Population Correction

The finite population correction (fpc) is an adjustment factor that becomes part of the formula when sample elements are selected without replacement.[8] However, the fpc has very little effect on the end result when the size of the sample is less than 5% of the total population. In most sample surveys, this is the case, and the fpc is excluded. In the following examples, we exclude the fpc; we have more to say about the relationship between sample size and population size in a later example.

Probability Level

The formula now has three components. To solve the equation, we set the values for two components (probability level and confidence interval) and we approximate the third (variance). The value for t, **probability level**, is the standard deviation score that expresses the percentage of a variable's values that fall within a set interval when the variable is normally distributed. One standard deviation includes approximately 68% of the sample values and its score is 1.0; two standard deviations include approximately 95% of the sample values and its score is 1.96; and three standard deviations include approximately 99% of the values and its score is 2.58. By using these values or scores, we assume our variable is normally distributed. When we set t, we are setting a probability level for our sample result. We are saying that out of 100 samples of the size that we are going to conduct, we want to be x percent confident (we set x) that our sample confidence interval (d) includes the population value.

Variance

The formula is for a variable expressed as a percentage. Typically this means that we are interested in a variable expressed as two categories: those who do and those who do not or those who have and those who do not have. For example, we may want to estimate the percentage of the population that smokes cigarettes, owns a handgun, rides public transportation, makes more than $30,000 per year, and so forth. Each of these variables is expressed as the percentage that do or have and the percentage that do not or do not have. Any variable can be expressed as a percentage. We may be interested in a variable that has response categories of strongly agree, agree, disagree, and strongly disagree. We can dichotomize this variable by deciding what category or categories we are interested in determining the sample size for. If we were interested in the proportion that agree with a statement, we would combine into one category those that strongly agree and those that agree and combine the two disagree categories into a second category. The percentages in the final two categories must sum to 1.0. The variance of a percentage variable is the product of the two percentages.

If we want to do a study to estimate the number of adult smokers in Raleigh, to determine the sample size, we need to come up with a good guess or estimate of the percentage of smokers before we do the study.[9] This estimate is a necessary component of the formula. While this seems incongruous with our task, there are a number of methods that can be used to make this estimate. One is to use the results from a previous study. If the characteristics of the study site are not similar to those of Raleigh, the researcher can adjust the estimate. Another method is to make a conservative guess at the proportion of smokers. As we show shortly, the variance component does not affect the solution of the formula as much as the confidence interval. Thus, an incorrect guess does not affect the required sample size as much as the setting of the precision of a confidence interval around the estimate. A third method for estimating the variance is to conduct a small pilot study. Depending on one's resources, a short phone call to a random sample of 25 to 50 households should provide a reasonable estimate. In this example, p would be the proportion of adults who smoke and q would be 1 minus that percentage.

Confidence Interval

The third component of the formula, which is determined by the researcher, is the confidence interval, or d. This interval is the margin of error that we require or will tolerate. The only way, theoretically, to estimate a population value without sampling error is to include or interview every element in the population. When we interview a sample of cases, we can only estimate the population value within a range of values. The component d is expressed as plus and minus and represents one-half of the range. Thus, if we assume that 30% of the adults in Raleigh smoke, d is how large or small we need the estimated range to be around 30%. As we show shortly, the smaller the range, the larger the required sample size.

Let's go back over the components of the formula and then go through an example. The variance is the value we must guess or estimate. The percentage of the population with the characteristic is p, and the remaining percentage of the population is q. The probability level, t, which we select, represents in how many repeated samples of 100 that total the same size as ours the population value is likely to fall within the specified confidence interval. The confidence interval, d, is the width of the range in which we want our estimate to fall.

Putting It All Together

Let's finish the example on estimating the number of adult smokers in the city of Raleigh. The question is, "What sample size do we need *if we assume the following components?*" Let's assume that Raleigh is similar to the U.S. population in the proportion of adults who smoke. Our assumptions are that Raleigh is the home of a major university and that its residents have an above-average level of education, but it is also the capital city of a tobacco-growing state. We assume these factors are offsetting and believe the percentage of adult smokers in the United States is a good estimate for Raleigh. We assume $p = .30$ and, therefore, q is $1 - .30 = .70$. We want our probability or confidence level to be 95%; that is, to include the population value in 95 of every group of 100 samples of the same size. And we want the confidence interval to be ±5%. In summary, the problem is stated as follows: "What sample size do we need if we want to be 95% confident that the population value lies in the interval between 25% and 35%?" Substituting these values into the formula gives

$$n = \frac{(1.96)^2(.30)(.70)}{(.05)^2} = \frac{.8067}{.0025} = 323$$

The number of interviews we need is 323. Remember that our estimate of the variance was a guess or a rough estimate. If we do 323 interviews and find that our estimate is wrong, what effect does that have on our calculations? The answer is that the confidence interval will be affected. The purpose for doing these calculations is to get some idea of the required sample size within the constraints we have established for the research problem. But these calculations are only a guide. We don't need to do exactly 323 interviews. However, if our estimate of variance is correct and we want to achieve the stated level of precision, we should strive to complete about this number of interviews.

Once the study is conducted, we always compute confidence intervals for the key variables. The confidence intervals are based on the variability of each variable, the number of completed interviews, and the probability level we established. In the above example, if our estimate of the proportion of smokers was too high, meaning less variance in the population, our final confidence interval would be smaller than ±5%. This would give our sample estimate more precision than we required. If our estimate was too low, meaning there are more smokers than we estimated, it would make our confidence interval larger than we planned for. Two examples illustrate this. We use the same formula as before, but this time we solve for d. We use the actual percentages or variances in the data and the number of interviews completed. If we found that 20% of the population smoked, the results for d would be

$$d^2 = \frac{(1.96)^2(.2)(.8)}{323} \qquad d = \sqrt{.00190} = .0436$$

The confidence interval would be ±4.4% around the sample estimate rather than ±5%. If our survey found 40% of the adults to be smokers, d would be

$$d^2 = \frac{(1.96)^2(.4)(.6)}{323} \qquad d = \sqrt{.00285} = .0534$$

Rather than ±5%, the confidence interval would be ±5.3%.

These examples illustrate an additional point. The formula has four components. By setting or knowing three of the components, you can solve the equation for the fourth.

Formula Assumptions

We should also point out other important aspects of the formula. It is based on selection of elements by simple random sampling (srs), which means that all elements and combinations of elements have an equal chance of selection. In reality, few samples are srs; thus, the formula is only a guide to sample size requirements. Second, the formula applies only to a variable in percentage form. Metric variables such as income require the more traditional method of calculating a variance, which is taught in data analysis or statistics classes. Third, the formula gives a sample size solution for only one variable. Most surveys have many important variables. The fact that each variable in a survey has its own variance and, thus, that each can require different sample sizes, forces researchers to make tough decisions. Typically, a compromise is fashioned between sample size requirements, the method of data collection, and the resources available.

A fourth assumption of the formula is that the sample comes from a large population or that the sample is a small proportion of the population. When this occurs, you can ignore the fpc, which we did in the above solutions. Many times, researchers incorrectly believe that a sample must be a certain percentage of the population if it is to accurately reflect the population. Sometimes we hear researchers ask if they should take a 1%, 5%, or some other percentage sample of the population. The only time the population size is important is when the sample represents more than 5% of the population, an arbitrary standard used by many sampling statisticians. Let's illustrate this with an example. Our sample size requirement for the smoking study is 323. The population correction factor is represented by (Cochran, 1977)

$$n' = \frac{n}{1 + \left(\frac{n-1}{N}\right)}$$

Our sample size requirement is $n = 323$ and the population of adults 18 years and older in Raleigh in 2000 is $N = 218,487$ (U.S. Census Bureau, 2000d). Substituting these values into the equation, we get

$$n' = \frac{323}{1 + \dfrac{322}{218,487}} = \frac{323}{1.00147} = 322.5$$

The solution gives us a difference of less than 1, which is really no difference. What might happen if the population of adults numbered 50,000? Substituting this number into the equation for N gives us a result of 321. Again, no real difference. Let's do one more example. What if the size of the population is 2,500? Substituting this into the equation gives a sample size requirement of 286. This represents 37 fewer interviews, or an 11.5% reduction. Note that a sample of 323 is 12.9% of a population of 2,500, which exceeds the 5% standard. This returns us to an earlier point: Population size does not affect sample size unless the population is small and the sample is more than 5% of the population. This is the reason why, if the variance and the other components of the formula are assumed to be the same, a study done in Raleigh or in the entire state of North Carolina or in the entire United States, requires the same size sample!

What Most Affects Sample Size

Now that we have an idea of how to determine sample size and we know that population size is usually not a factor, we want to examine what component or components of the formula have the greatest effect on sample size. Exhibit 7.1 presents the sample sizes for different combinations of variances, probability levels, and confidence intervals. Part A presents the results for 95% confidence (probability level), and Part B for 90% confidence (probability level). The results for a third, frequently used, probability level, 99%, is left for the student to solve. Within each part of the table, we present the maximum variance for a percentage variable (.50/.50), a middle range variance (.70/.30) and a low estimate of variance (.90/.10). We also present three confidence intervals: ±5%, ±4%, and ±2%.

Let's first look at Part A, which gives the 95% probability level. For a variable with the maximum variance of .50/.50 and a confidence interval of ±5%, the sample size requirement is 384. Going down the first column, the

Exhibit 7.1   Required Sample Sizes for Different Variances (expressed as percentages), Probability Levels, and Confidence intervals[a]

A. 95% Probability level with different variances and confidences intervals.

$$\frac{(1.96)^2(.5)(.5)}{(.05)^2} = 384 \qquad \frac{(1.96)^2(.5)(.5)}{(.04)^2} = 600 \qquad \frac{(1.96)^2(.5)(.5)}{(.02)^2} = 2401$$

$$\frac{(1.96)^2(.7)(.3)}{(.05)^2} = 323 \qquad \frac{(1.96)^2(.7)(.3)}{(.04)^2} = 504 \qquad \frac{(1.96)^2(.7)(.3)}{(.02)^2} = 2017$$

$$\frac{(1.96)^2(.9)(.1)}{(.05)^2} = 138 \qquad \frac{(1.96)^2(.9)(.1)}{(.04)^2} = 216 \qquad \frac{(1.96)^2(.9)(.1)}{(.02)^2} = 864$$

B. 90% Probability level with different variances and confidences intervals.

$$\frac{(1.64)^2(.5)(.5)}{(.05)^2} = 269 \qquad \frac{(1.64)^2(.5)(.5)}{(.04)^2} = 420 \qquad \frac{(1.64)^2(.5)(.5)}{(.02)^2} = 1681$$

$$\frac{(1.64)^2(.7)(.3)}{(.05)^2} = 226 \qquad \frac{(1.64)^2(.7)(.3)}{(.04)^2} = 353 \qquad \frac{(1.64)^2(.7)(.3)}{(.02)^2} = 1412$$

$$\frac{(1.64)^2(.9)(.1)}{(.05)^2} = 98 \qquad \frac{(1.64)^2(.9)(.1)}{(.04)^2} = 151 \qquad \frac{(1.64)^2(.9)(.1)}{(.02)^2} = 605$$

[a]These calculations do not include the finite population correction factor.

requirements for a variable with a variance of .70/.30 is 323, and for one with a variance of .90/.10, 138. Note that the smaller the variance, the smaller the required sample size.

In each row, the confidence intervals change. For the maximum variance and for a confidence interval of ±5%, the sample size required is 384. As we demand more and more precision around the estimate, the required sample size increases. At ±4% the requirement is 600, but at ±2% the requirement jumps to a sample size of 2,401. Going from ±5% to ±2% causes more than a sixfold increase in sample size. This increase is the same for each level of variance. When we go from a low variance to a maximum variance, the required sample size increases by a factor of 2.77; however, when we go from a moderate confidence interval to a small confidence interval, the increase is a factor of 6.25. These changes are comparable across all variances, confidence intervals, and the two probability levels. Thus, for a percentage variable, the size of the confidence interval or the precision of the sample estimate affects sample size requirements the most. The Current Population Survey is a real-life example of this requirement for a metric variable that must be estimated with a high level of precision. This survey is conducted monthly to estimate the unemployment rate in the United States with a confidence interval that is a fraction of 1%. To achieve this percentage rate, approximately 56,000 households are sampled each month.

Hypothesis Testing and Power

Our discussion thus far has focused only on the rudimentary components of determining sample size. Rather than estimating a population percentage, we are frequently interested in testing hypotheses or determining sample size and power. The reader may find a few more elementary examples useful if we conceptualize the research questions from these different perspectives. Let's do a few examples of hypotheses testing. Assume that a number of recent national surveys indicate that approximately 43% of registered voters believe that Congress is doing a good job. An investigator wants to compare the attitudes of voters in her home state to those nationally. She interviews a sample of 200 currently registered voters and finds that 51% in her state believe that Congress is doing a good job. She wants to test the hypothesis that her state is similar to the national voters. She assumes a nondirectional hypothesis (H_0: $p = .43 = p_0$ or H_1: $p \neq .43$), uses the U.S. results as the criterion, sets $\alpha = .05$, assumes the normal approximation to the binomial distribution is valid, and tests the differences by converting the p into a z score:

$$z = \frac{|\hat{p} - p_0| - 1/(2n)}{\sqrt{\dfrac{p_0 q_0}{n}}}$$

Where

$\hat{p}$ = *sample estimate*

p_0 = *population value*

$q_0 = 1 - p_0$

$\dfrac{1}{2n}$ = *correction for continuity for the binomial*

$$z = \frac{|.51 - .43| - .0025}{\sqrt{\dfrac{(.43)(.57)}{200}}} = 2.21$$

Because $z > 1.96$ we can reject H_0 and conclude that the state voters feel differently than the national voters. If the sample size had been n = 100 and the investigator got the same approval results, the result for z would have been z = 1.57. In this situation, we would not have rejected H_0. These different results are a consequence of sample size and this brings up the issue of power, which we deal with after the next example.

The next example involves testing a hypothesis between two independent random samples. Assume that a sample of n = 300 individuals are interviewed in each of two towns. In town one, 52% of the respondents say they are Democrats and in town two, 45% of the respondents say they are Democrats. We want to determine if this difference is significant or a result of chance. As before we test the hypothesis H_0: $p_1 = p_2 = p$ vs. H_1: $p_1 \neq p_2$. We assume the normal approximation to the binomial is valid, a nondirectional hypothesis, and set $\alpha = .05$. The formula is

$$z = \frac{|\hat{p}_1 - \hat{p}_2| - \left(\dfrac{1}{2n_1} + \dfrac{1}{2n_2}\right)}{\sqrt{\hat{p}\,\hat{q}\left(\dfrac{1}{n_1} + \dfrac{1}{n_2}\right)}}$$

Where

$\hat{p}_1$ = is the estimate from town one

$\hat{p}_2$ = is the estimate from town two

$\hat{p}$ = is a weighted average of the results from the two towns

$$\hat{p} = \frac{n_1\hat{p}_1 + n_2\hat{p}_2}{n_1 + n_2}$$

$$\hat{q} = 1 - \hat{p}$$

Substituting in our results gives

$$\frac{|.52 - .45| - \left(\dfrac{1}{600} + \dfrac{1}{600}\right)}{\sqrt{(.485)(.515)}\left(\dfrac{1}{300} + \dfrac{1}{300}\right)} = \frac{.07 - .00333}{\sqrt{.001665}} = 1.63$$

Because z is less than 1.96 we would not reject H_0 and we would conclude that the proportion of Democrats in the two towns may be similar.

Interestingly, if we were to increase the sample sizes to n = 400 in each town and we were to get the same town proportions of Democrats, the results would be borderline significant, depending on how we did the calculations. Using n = 400, the result would be

$$z = \frac{|.52 - .45| - \left(\dfrac{1}{800} + \dfrac{1}{800}\right)}{\sqrt{(.485)(.515)\left(\dfrac{1}{400} + \dfrac{1}{400}\right)}} = \frac{.0675}{.03534} = 1.91$$

Because this value is less than 1.96, we would not reject H_0. However, if we were to exclude the correction for continuity in the numerator, the result would be z = .07/.03534 = 1.98. In this instance we would reject H_0 and conclude that the two towns are different. This very nicely leads into our discussion and examples on power.

When we test hypotheses, we usually formulate a null hypothesis (H_0) and test it against an alternative hypothesis (H_1). Most statistics texts outline in detail the steps and assumptions required to perform hypothesis testing (see, e.g., Glenberg, 1996). When you test a hypothesis, four possibilities can occur:

1. Accept H_0 when it is correct,

2. Accept H_1 when H_0 is true (Type I error),

3. Accept H_0 when H_1 is true (Type II error), or

4. Accept H_1 when it is correct.

Thus, there are two types of errors that we want to guard against. The probability of a Type I error, α, is the probability of rejecting H_0 when it is true and should *not* be rejected. In the social sciences we typically set α, or the probability of this error, at .05. This means that the probability of making this error is 5% and the probability of being correct when we test an hypothesis is $1 - \alpha$, or 95%. So the probability of this type of error is small. We can, however, set the probability much lower, like $\alpha = .025$ or .01, but it has consequences for the second type of error.

The probability of a Type II error, β, is the probability of accepting H_0 when H_1 is correct and should be accepted. β and the power of a test are related. Power, which is $1 - \beta$, is the probability of rejecting H_0 when it should be rejected. The two types of errors, α and β, are inversely related. By making α small, we increase β, and by reducing β, we increase α. Power is important because low power means that there is a small chance of finding a significant difference between variables when a real difference may exist. So why is it important to give this much attention to α, β, and power? The main reason is that small sample sizes produce tests with low power and power analyses should be conducted as part of the sample size determination. The purpose of power calculations is to plan a sample size that maximizes the probability of rejecting H_0 when it should be rejected.

Although there are a number of ways to affect power (see, e.g., Rosner, 2000; Glenberg, 1996), we want to illustrate only the effects caused by sample size. Let's return to our earlier example about registered voters and is Congress doing a good job. We will do a one sample binomial test with a two-tailed test at $\alpha = .05$ for $n = 200$. What level power do we have for rejecting H_0? Lachin (1981) provides the following single proportion formula

$$z_\beta = \frac{\sqrt{n}|\hat{p} - p_0| - z_{\alpha/2}\sqrt{p_0 q_0}}{\sqrt{\hat{p}\hat{q}}}$$

Substituting our values in, we get

$$z_\beta = \frac{\sqrt{200}\,|.51 - .43| - 1.96\sqrt{(.43)(.57)}}{\sqrt{(.51)(.49)}} = \frac{1.13 - .9703}{.499} = .322$$

We look up .322 in a table of values for the normal distribution (see Exhibit 7.2) where we find that $Z_\beta = .322$ lies between $Z_\beta = .39$ ($\beta = .35$, power = .65) and $Z_\beta = .25$ ($\beta = .40$, power = .60). So, a sample size of 200 yields a test with power between .60 and .65, or approximately 63% power. For exact values refer to a more detailed table or consult one of the sources listed in the last two paragraphs of this section. If we recalculate based on a sample size of n = 100, we get a negative result of −.34. If we look up this value in a normal distribution table and subtract it from 1, we find that a sample size of 100 yields 37% power. Therefore, there is only about a 1 in 3 chance of rejecting the null hypothesis or a 63% chance of making a Type II error for n = 100. In this example, a sample size of 100 is not a strong test of the hypothesis.

One convention that has evolved is to set β to be four times α. So if $\alpha = .05$, β would be .20 and power would be $1 - \beta = .80$. Using our previous assumptions, what sample size would be needed to have 80% power? We use

$$n = \left[\frac{z_{\alpha/2}\sqrt{p_0 q_0} + z_\beta\sqrt{\overline{p}\,\overline{q}}}{\overline{p} - p_0}\right]^2$$

Exhibit 7.2   Unit Normal Deviates for Selected Values of α and β

α or β	Z_α for one sided test or Z_β	$Z_{\alpha/2}$ for two sided test
.01	2.33	2.57
.02	2.05	2.33
.025	1.96	2.24
.04	1.75	2.05
.05	1.65	1.96
.10	1.28	1.65
.20	.84	1.28
.30	.52	1.04
.35	.39	.93
.40	.25	.84

Substituting our values in, we get

$$n = \left[\frac{1.96\sqrt{(.43)(.57)} + .84\sqrt{(.51)(.49)}}{.51 - .43} \right]^2 = \left[\frac{.9703 - .4199}{.08} \right]^2 = 302$$

Frequently we want to determine sample size and power for two independent samples. A typical situation might be if we were designing an intervention study. Suppose we wanted to reduce smoking by 30% in an adult population where 33% of the adults smoked. Assume that we want equal-sized control and experimental groups ($n_c = n_e$), $\rho_c = .33$, $\rho_e = .23$ ($= .70 \times .33$) and $\bar{p} = .28$ (null hypothesis which is the average of ρ_c and ρ_e). We will use a two-tailed test for $\alpha = .05$ and power equal to 80%. Lachin (1998) provides the following equation for total sample size:

$$n = \left[\frac{z_{\alpha/2}\sqrt{4\bar{p}\bar{q}} + z_{\beta}\sqrt{2\rho_e q_e + 2\rho_c q_c}}{|\rho_e - \rho_c|} \right]^2$$

Substituting in our values, we get

$$n = \left[\frac{1.96\sqrt{4(.28)(.72)} + .84\sqrt{2(.23)(.77) + 2(.33)(.67)}}{|.33 - .23|} \right]^2$$

$$= \left[\frac{1.76 + .7496}{.10} \right]^2 = 630$$

We would need 315 subjects in the experimental group and in the control group for a total sample size of 630. Let's assume that our funding only allows for a total sample of 500 cases. What level of power could we expect? Again, from Lachin (1998):

$$= \frac{\sqrt{n}\,|p_e - p_c| - z_{\alpha/2}\sqrt{4\bar{p}\bar{q}}}{\sqrt{2p_e q_e + 2p_c q_c}}$$

$$= \frac{\sqrt{500}\,|.33 - .23| - 1.96\sqrt{4(.28)(.72)}}{\sqrt{2(.23)(.77) + 2(.33)(.67)}}$$

$$= \frac{2.236 - 1.76}{.892} = .534$$

This value in the normal distribution table indicates 70% power.

There are a number of excellent sources that can be consulted for issues on sample size and power. The book by Arthur Glenberg (1996), *Learning from Data: An Introduction to Statistical Reasoning* provides a very basic explanation of power for one- and two-independent samples. *Fundamentals of Biostatistics* by Bernard Rosner (2000) is a more advanced text. This book has numerous health examples covering cancer, cardiology, hypertension, obstetrics, and other areas. Examples cover hypothesis testing and confidence intervals, using one- or two-sided alternatives, for one, two, or longitudinal samples, and for binomial and metric variables when determining sample size and power.

There are a number of excellent Web sites. One site with more than 600 links was developed and is maintained by John C. Pezzullo, a retired biostatistician from Georgetown University, and can be found at http://members.aol.com/johnp71/javastat.html. This site includes links to choosing the right statistical procedure, free software, books and manuals, tutorials, and the calculation of numerous statistical procedures; it takes 14 pages to print all the links at this site. On the first page is a link for calculating power, sample size, and experimental design. Another very useful site, developed by Russell V. Lenth, can be found at http://www.cs.uiowa.edu/~rlenth/Power/. This site does calculations for power, sample size, and confidence intervals; provides advice on the calculation of power and sample size; and gives a few links to other useful sites.

Using Census Data

As pointed out frequently, census data can be helpful in planning a survey, especially for estimating how many interviews we will obtain with people from various demographic subgroups and whether these sample sizes will be adequate for purposes of analysis. To ensure that we have adequate sample sizes, we must plan ahead. We need to review our research questions again: What are the objectives of our survey? What are the questions that we want to answer? Do we need answers for the total population? Only for specific subgroups? If so, which ones? Or, do we need answers for both the total population and certain subgroups? Let's continue the discussion of our survey about adult smokers in Raleigh.

Research Objectives

We need to think about our research objectives and the types of analyses we feel are important. As we do this, we want to consult census data to see what the interview outcomes might be. Again, we begin with: What is the research question? We want to estimate the number of adult smokers in Raleigh because we want to develop and test an intervention program to see whether it is successful in helping people to quit smoking. A natural followup question is whether we need to consider specific types of people. Are we interested only in an estimate of the total number of smokers, or do we want to get estimates for specific subgroups defined by gender, age, race, or combined gender and race? We ask these questions because research tells us that smoking behavior differs among these subgroups and it will be more effective if we tailor the intervention programs to specific needs of certain subgroups, rather than developing one general program for everyone. Therefore, we are interested not only in total population estimates but also in estimates for specific subgroups.

Analysis Groups

The next step is to ask ourselves in what specific subgroups we are interested. Let's assume we are interested in estimates for both males and females; for three age groups: 18 to 39 years, 40 to 64 years, and 65 years and older; by race; and maybe by gender and race. Next, we need to find out how many people there are in Raleigh in these various groups, determine how many interviews we can afford to conduct, and then put these pieces of information together. This last step will allow us to estimate the confidence intervals for all the subgroups and to determine whether the levels of precision are adequate for our purposes.

Assume that our resources will allow us to conduct 600 thirty-minute telephone interviews. To estimate how these interviews will be divided among our subgroups of interest, we need to find data that describe the adults in Raleigh by the characteristics of interest, and we need to make assumptions about whether a telephone survey will adequately represent these groups.

Consulting Census Data

It is at this point that we consult census data. The U.S. Census Bureau conducts censuses on population and housing, agriculture, and government; economic censuses on retail trade, wholesale trade, service industries, transportation, manufactures, mineral industries, and construction industries; and censuses on foreign trade and other subjects. Censuses are taken every 5 or 10 years, depending upon the topic. Data are presented for two types of areas: governmental—United States, Puerto Rico, states, counties, cities, villages, townships, congressional districts, Indian reservations, and so forth; and statistical areas—regions of the United States, metropolitan and micropolitan areas, urban areas, census tracts, blocks or block groups, ZIP codes, and other types of geographic areas. Data are available on line at http://factfinder. census.gov, in printed reports, on CD-ROMs, and on microfiche. In addition to these censuses, the Census Bureau conducts about 250 sample surveys each year. In short, the U.S. Census Bureau gathers and disseminates a treasure trove of information.

For our survey on smokers in Raleigh, we are interested in information that comes from the census of population and housing. Conducted every 10 years, this census collects data from every housing unit. A short-form questionnaire includes questions about the sex, race, age, marital status, and household relationship of all residents of the unit. The long-form questionnaire, which is distributed to a sample of people, includes additional questions, on social and economic characteristics such as education, place of birth, ancestry, language spoken at home, disability, fertility, employment characteristics, and other topics.

The information we need can be found at http://factfinder.census.gov using data from Census 2000 Summary File 1 (SF 1). Exhibit 7.3 provides data about the total number of people in the city of Raleigh by specific ages and age groups, by race and Hispanic origin, and by one gender group—females. To determine the number of males by age and race we need to subtract the number of females from the total number of people. By doing a little addition, subtraction, and division, we can obtain from this table all the information that we require.

Making Choices

Exhibit 7.4 provides an estimate of the number of completed interviews, by subgroup, for a sample survey of 600. Before constructing this table, we first consulted the Census of Housing (U.S. Census Bureau, 2000a) and learned that only 1.6% of the occupied housing units in Raleigh do not have a telephone. Therefore, we are assuming that telephone coverage is fairly uniformly distributed among the subgroups in the population and that none of our subgroups of interest are significantly underrepresented. We are also assuming that a telephone survey will reach each of these subgroups in the same proportions in which they are distributed in the population; that is, if our sample is truly random and we do not get differential rates of interview cooperation by subgroups, the final results of our survey by subgroups should closely approximate the percentages in the population. We don't expect our numbers to exactly match those in Exhibit 7.4, but we assume that they will be reasonably close to these estimates.

Exhibit 7.3   2000 Population of Raleigh by Age, Sex, Race, and Hispanic Origin

Raleigh City

Place and [in Selected States] County Subdivision [10,000 or More Persons]	All persons	White	Black	American Indian, Eskimo, or Aleut	Asian or Pacific Islander	Hispanic origin (of any race)	White, not of Hispanic origin
All persons	276,093	178,649	78,844	2,091	10,872	19,308	166,386
Under 5 years	17,461	9,633	6,469	142	788	2,043	8,177
Under 1 year	3,768	2,158	1,309	36	146	523	1,798
1 year	3,653	2,021	1,329	24	195	445	1,698
2 year	3,433	1,867	1,306	26	140	407	1,590
3 year	3,328	1,825	1,236	26	150	354	1,570
4 year	3,279	1,762	1,289	30	157	314	1,521
5 to 9 years	16,444	8,711	6,702	127	683	1,304	7,726
5 years	3,314	1,785	1,310	21	130	315	1,544
6 years	3,348	1,768	1,383	25	140	271	1,558
7 years	3,255	1,698	1,329	29	142	264	1,514
8 years	3,299	1,746	1,354	16	144	236	1,565
9 years	3,228	1,714	1,326	36	127	218	1,545
10 to 14 years	15,254	8,405	5,963	128	616	903	7,725
10 years	3,290	1,780	1,325	24	129	189	1,632
11 years	3,191	1,747	1,267	24	130	198	1,594
12 years	3,030	1,669	1,184	28	123	171	1,541
13 years	2,865	1,605	1,098	25	104	190	1,474
14 years	2,878	1,604	1,089	27	130	155	1,484
15 to 19 years	19,864	12,355	5,934	187	801	1,924	11,169
15 years	2,805	1,613	1,007	19	130	179	1,478
16 years	2,736	1,565	989	29	97	254	1,391
17 years	2,906	1,649	988	32	137	353	1,413
18 years	4,772	3,038	1,334	51	183	513	2,742
19 years	6,645	4,490	1,616	56	254	625	4,145

Raleigh City

Place and [in Selected States] County Subdivision [10,000 or More Persons]	All persons	White	Black	American Indian, Eskimo, or Aleut	Asian or Pacific Islander	Hispanic origin (of any race)	White, not of Hispanic origin
20 to 24 years	32,458	21,072	8,213	308	1,437	3,763	19,036
20 years	6,725	4,545	1,590	64	239	706	4,173
21 years	6,616	4,351	1,672	72	277	741	3,926
22 years	6,538	4,267	1,667	56	281	741	3,845
23 years	6,340	3,998	1,668	59	311	786	3,590
24 years	6,239	3,911	1,616	57	329	789	3,502
25 to 29 years	30,850	19,591	8,138	246	1,568	3,281	17,763
30 to 34 years	26,255	16,436	7,588	210	1,226	2,251	24,004
35 to 39 years	23,320	14,866	6,908	185	939	1,495	21,825
40 to 44 years	20,506	13,334	5,976	170	821	885	19,621
45 to 49 years	17,950	12,240	4,853	108	658	571	17,379
50 to 54 years	15,034	10,722	3,694	97	461	359	14,675
55 to 59 years	10,308	7,657	2,282	48	307	204	10,104
60 to 64 years	7,394	5,561	1,605	40	195	121	7,273
65 to 69 years	6,246	4,707	1,350	26	171	75	6,171
70 to 74 years	5,779	4,522	1,150	24	90	52	5,727
75 to 79 years	4,957	3,939	947	15	63	39	4,918
80 to 84 years	3,186	2,600	556	20	29	21	3,165
85 years and over	2,827	2,298	516	10	19	17	2,810
18 years and over	218,487	147,073	56,726	1,614	8,421	14,272	52,570
62 years and over	27,249	21,275	5,442	115	482	259	26,990
65 years and over	22,995	18,066	4,519	95	372	204	22,791

(Continued)

159

Exhibit 7.3 (Continued)

	Raleigh City						
Place and [in Selected States] County Subdivision [10,000 or More Persons]	All persons	White	Black	American Indian, Eskimo, or Aleut	Asian or Pacific Islander	Hispanic origin (of any race)	White, not of Hispanic origin
Median age	30.9	32.7	28.8	28	28.6	24.6	31.8
Female	139,445	89,443	42,219	1,019	5,183	7,339	84,392
Under 5 years	8,548	4,668	3,224	69	391	977	3,973
Under 1 year	1,845	1,044	658	19	74	256	863
1 year	1,798	990	666	11	98	211	828
2 year	1,678	912	642	12	74	187	786
3 year	1,617	867	612	11	71	177	751
4 year	1,610	855	646	16	74	146	745
5 to 9 years	8,064	4,271	3,278	68	355	627	3,792
5 years	1,604	848	643	14	78	142	731
6 years	1,653	854	689	11	75	149	741
7 years	1,617	849	651	12	77	130	758
8 years	1,589	848	650	10	60	111	770
9 years	1,601	872	645	21	65	95	792
10 to 14 years	7,563	4,095	3,020	75	310	448	3,759
10 years	1,627	861	682	14	61	95	778
11 years	1,550	828	640	12	64	85	760
12 years	1,459	803	567	16	59	87	745
13 years	1,467	822	557	16	64	101	754
14 years	1,460	781	574	17	72	80	722
15 to 19 years	9,798	6,005	3,137	84	368	726	5,557
15 years	1,382	782	519	10	63	84	713
16 years	1,342	751	506	11	50	104	680
17 years	1,408	786	506	14	67	119	706
18 years	2,410	1,535	721	23	78	183	1,430
19 years	3,256	2,151	885	26	110	236	2,028

Raleigh City

Place and [in Selected States] County Subdivision [10,000 or More Persons]	All persons	White	Black	American Indian, Eskimo, or Aleut	Asian or Pacific Islander	Hispanic origin (of any race)	White, not of Hispanic origin
20 to 24 years	15,474	9,923	4,410	149	625	1,241	9,182
20 years	3,235	2,173	853	28	103	249	2,028
21 years	3,106	2,006	884	35	130	232	1,855
25 to 29 years	14,617	9,073	4,354	100	734	1,063	8,417
30 to 34 years	12,672	7,740	4,102	83	573	757	11,915
35 to 39 years	11,434	7,153	3,655	94	431	520	10,914
40 to 44 years	10,399	6,625	3,228	86	409	335	10,064
45 to 49 years	9,342	6,251	2,688	60	345	217	9,125
50 to 54 years	7,901	5,613	2,007	43	227	162	7,739
55 to 59 years	5,412	3,970	1,265	26	148	92	5,320
60 to 64 years	4,015	3,001	899	25	95	65	3,950
65 to 69 years	3,491	2,591	804	15	88	40	3,451
70 to 74 years	3,395	2,613	732	8	44	29	3,366
75 to 79 years	3,012	2,354	623	11	32	17	2,995
80 to 84 years	2,138	1,741	384	17	13	11	2,127
85 years and over	2,170	1,756	409	6	13	12	2,158
18 years and over	111,138	74,090	31,166	772	3,947	4,980	106,158
62 years and over	16,490	12,757	6,470	72	245	137	16,353
65 years and over	14,206	11,055	2,952	57	190	109	14,097
Median age	32.1	34.2	29.6	27.9	28.7	23.6	32.8
Male	136,648	89,206	36,625	1,072	5,689	11,969	124,679
Median age	30.0	31.4	27.7	28.1	28.3	25.1	30.8
Males per 100 females	98.0	99.7	86.8	105.2	109.8	163.1	94.4

Exhibit 7.4   Estimates of the Number of Telephone Interviews for Selected Subgroup Characteristics for a Survey of 600 Adults in the City of Raleigh

| | Population | | |
	N	%	Interview Estimate
Race			
White	147,073	67.3	404
Black	56,726	26.0	156
Other	14,688	6.7	40
	218,487	100.0	600
Gender			
Male	107,349	49.1	295
Female	111,138	50.9	305
Age			
18–39	124,300	56.9	341
40–64	71,192	32.6	196
65 and older	22,995	10.5	63
18–34	100,980	46.2	277
35–64	94,512	43.3	260
65 and older	22,995	10.5	63
18–34	100,980	46.2	277
35–59	87,118	39.9	239
60 and older	30,389	13.9	83
Race and Gender			
White males	72,983	33.4	200
White females	74,090	33.9	203
Black males	25,560	11.7	70
Black females	31,166	14.3	86

Exhibit 7.4 shows a common situation for gender in that the population is fairly evenly divided between males and females. A survey should give us approximately equal sample sizes by gender if cooperation rates are the same. For most surveys, however, this is a dubious assumption because females are more likely to cooperate than males. Final interview results for males are usually a few percentage points below their population proportion. If males constitute 49% of the eligible population, as they do in this example, they may represent between 42% and 46% of the final interviews. At the moment, let's not worry about adjusting the numbers in the table by guessing the cooperation rates. Let's use the numbers in the table as our best guess of the outcomes by subgroup. The question we need to ask is, "Is the number of interviews to be conducted with each of these subgroups adequate to meet our objectives?" In effect, can we live with the levels of precision the various sample sizes will produce?

Let's do a few calculations and see. For males we expect about 295 interviews. If we set a 95% probability and we assume a .30/.70 variance, and solve for d, we get a result of ±5.2%. For what we are trying to do, a confidence interval (range) of 10.4% is reasonable.

Let's look at race. Our survey should yield 404 interviews with whites, 156 with blacks, and only 40 interviews with people of other races. There are two major inequities here. One, 40 interviews with any diverse group are too few for any meaningful analysis. Our choices are to (a) combine people of other races with whites or blacks; (b) screen other races out of the study by asking about the respondent's race at the start of the survey and treating people who are not white or black as ineligibles; or (c) interview all respondents and exclude other races from the analysis. Choices b and c seem best, but this is not an easy decision. The strategy of combining other races with the whites or blacks is unacceptable. It is very likely, and the literature seems to support this, that the attitudes and smoking behaviors of people of Chinese, Japanese, Malaysian, and other descents might be very different from each other and from those who are white or black. Thus, there will be too few interviews with any one group for meaningful analysis. Choice b is risky. To open the interview with a personal question about race in order to avoid 40 ineligible interviews may not be worth the risk of offending potential respondents and increasing the overall refusal rate. With choice c, it is usually a bad strategy to throw away interviews, especially when one has a limited budget. The best strategy may be to test choice b while pretesting. If screening for race affects the refusal rate and/or affects costs, then choice c would be the best compromise strategy.

The second inequity concerns the great difference in the number of interviews that will be conducted with white and black respondents. Our previous sample size examples illustrate that when the variances between groups are equal but the sample sizes are different, the size of the confidence interval will also be different. Ideally, when comparing groups, we want the confidence intervals to be equal. Using the sample size estimates in Exhibit 7.4 and the assumptions used for the other races, the confidence intervals would be ±4.5% for whites and ±7.2% for blacks. The size of the interval range is 9% for whites and 14.4% for blacks. For most social science variables these ranges may be adequate; for others, however, more precision may be needed. We must remember that these ranges are for a .30/.70 variable split. As the percentages move toward .50/.50, the size of these ranges will increase. One alternative is to oversample blacks. This means screening households for race, which, as we mentioned earlier, may have negative consequences. Screening would also increase costs, reducing the number of interviews we could afford. The choices are not easy. In a situation like this, the researcher must look carefully at the key variables and the required degree of precision in making final design decisions.

Recall that each variable or combination of variables has its own variance. If we are interested in the attitudes and behavior of white males and females and black males and females, we need to do the same calculations. Exhibit 7.4 shows that we can expect about 200 interviews with each white gender group, but we will only have about 70 to 86 interviews with the black gender groups. We can get a rough idea of the confidence intervals for whites from our previous calculations, so let's see what it might be for black males, the smallest group. Using a .30/.70 variance estimate and a 95% probability level, a sample size of 70 gives a confidence interval of ±10.7% or a range of 21.4%. If our study found 30% of the black males to be smokers, the 95% confidence interval would provide a population estimate of 19.3% to 40.7% of the black males to be smokers. If this is a critical variable, we would want to consider the options for increasing this sample size. One additional comment. If we are interested in the behavior of young black males, we would get even fewer interviews and the sample size would be wholly inadequate.

The age figures in Exhibit 7.4 illustrate another important point. Initially, we had an interest in three age groups: 18 to 39 years (young), 40 to 64 years (middle), and 65 years and older (older). These groups are not equally distributed in the population. The bulk of the interviews will be with young adults, and the older adults have a smaller sample yield than even the black males. One additional option to better equalize the size of the confidence intervals is to change the definition of our age groups. We would do this only if it makes sense substantively. For example, we could define the young as 18 to 34 years of age and the middle age group as 35 to 64 years of age. This better equalizes each group's confidence intervals but it does not change the situation for the older respondents. Another possibility is to redefine the middle age group as 35 to 59 years of age and the older age as 60 years of age and older. This change gives us a slightly better sample size for the older adults than the sample size that we will get for the black males group. Other definitions are also possible.

We conclude by emphasizing that data are available to assist the researcher in planning for the most effective study possible. Even with a good research idea, one should not blindly march ahead and begin collecting data without planning to achieve the best possible outcome. It is unpardonable, for example, to decide to study attitudes of whites and African Americans, conduct the study, and then learn that African Americans comprise only 12% of the population and that the number of interviews with them is inadequate for data analysis. The elderly—those age 65 years and older—typically comprise a similar percentage of the total population. Before designing a study of the elderly, the researcher should know not only this proportion but also that the elderly usually have lower response rates than other segments of the population.

Information is available from census data, existing studies of similar populations, textbooks, data archives, and other sources to help design and plan an effective study. Chapter 10 and the References also provide a number of sources that should be consulted in designing or implementing a study.

Notes

1. An additional drawback to decennial census data is that the data become outdated after a few years, especially in high growth areas. One strategy to overcome this and enumerator difficulties is the proposed American Community Survey. This survey, conducted annually, will collect detailed demographic, socioeconomic, and housing data. It will provide annual and multi-year estimates of these characteristics and will ultimately replace the decennial census long form. Data quality comparisons between Census 2000 results and a large supplemental survey conducted at the same time can be found in U. S. Census Bureau (2004).

2. Sampling error is not error in the sense that we have made a mistake. It is an estimated range of values, computed from our sample, that indicates where the population value may be. The estimated range of values is computed because only a sample of the population, not everyone, is interviewed.

3. The reasons for conducting five interviews per block are discussed in Chapter 8.

4. Selecting places with probabilities proportionate to a measure of size in the first stage of sample selection gives larger places a higher chance of being in the sample. For example, we propose to select 24 counties from a total of 100 counties. The population of persons age 18 years and older in North Carolina in 2000 was 6,085,266. We create a sampling interval by dividing 6,085,266 by 24, which gives 253,553. Onslow County had 111,017 persons age 18 years and older, and Ashe County had 19,557. The probability of being selected in the first stage of selection is higher for Onslow (111,017/253,553 = .438) than for Ashe (19,557/253,553 = .077). To equalize the overall probability of selection, the same number of units must be selected at the last stage of selection. This point is illustrated in Chapter 8.

5. Weighting is discussed later in this chapter and in Chapters 8 and 10.

6. These points are illustrated in Chapter 8.

7. This information was obtained through a telephone call to the Industrial Analysis Division at the FCC. It does not include households that have both residential and business telephone numbers.

8. Sample elements can be selected with or without replacement. With replacement works as follows: If we want to randomly select a sample of three numbers from the set of numbers 1 to 10, after each number is selected, it is returned to the group of numbers and it can be selected again at the next selection of a number.

9. While we work through the example step-by-step, the reader might want to think through the steps using a different example, such as determining the sample size to estimate the proportion of adults in your home town that eat breakfast at least five times a week or who exercise for 30 minutes at least three times per week. These alternate examples will help to reinforce the step-by-step procedures.

8

Selecting a Sample

Telephone, mail, and Internet surveys are the three types of studies students are most likely to conduct. This chapter describes many of the decisions that must be made when selecting samples for a community telephone survey and how these procedures can be extended to select a national random-digit dialing (RDD) sample. We then illustrate two examples of selecting samples from lists: (a) where the order of the individuals can be manipulated to take advantage of the benefits of stratifying a sample, and (b) a sample of natural clusters such as classes of students.

Exhibit 8.1 outlines the major steps in the selection of a sample. The examples in this chapter illustrate these steps. Before you begin Example 1, you may want to read these steps so you can get a sense of how the discussion of the examples is organized.

Example 1: A Community List-Assisted Telephone Sample

Defining the Population

To begin, we review the early stages of a study. Assume we want to study attitudes of adults age 18 years and older in the Raleigh area about abortion. We have concluded that a telephone study is an acceptable method for collecting these data because this method has low response effects, and it fits

Exhibit 8.1   Steps in Sample Selection

- Define the population
- Find and/or develop a frame
- Determine sample size
- Select sample
- Manage sample and resample
- Select respondents

our time schedule and budget. **Response effects** are the types of error in the answers to questions that are a result of factors such as faulty memory, the respondent misunderstanding the question, or method of data collection (Sudman & Bradburn, 1982). We have also considered what proportion of households do not have telephones and we are willing to accept this potential bias. In Wake County, the estimate is 1.2%, and in the city of Raleigh, it is 1.6% (U.S. Census Bureau, 2000a). Both estimates are below the national average.

Selecting a Frame

Our next task is to find an acceptable sampling frame and to assess its coverage and recency. The Raleigh telephone directory has a residence section that presumably contains only households, no businesses. The directory includes Raleigh (2000 population = 276,093), Cary (2000 population = 94,536), six smaller cities (2000 population = 57,428), and rural, unincorporated areas (2000 population = 87,264). Most of the county population is covered by the directory. Because liberal and conservative attitudes about abortion may be correlated with location of residence, and we wish to test this and other hypotheses, the coverage of the telephone directory suits our purposes.

As a frame, however, the telephone directory has one major drawback. It does not include people with unlisted telephone numbers. An estimate by Survey Sampling International (Piekarski, 1997), indicates that 21.8% of the households in the Raleigh-Durham-Chapel Hill metropolitan statistical area are unlisted. This potential bias is too large to ignore; thus, our survey requires an RDD sample.

The current RDD method preferred by most survey organizations and government agencies is the list-assisted method. The method was first proposed and tested as a stratified list-assisted design by Casady and Lepkowski (1993). Brick, Waksberg, Kulp, and Starer (1995) and Tucker, Lepkowski, and Piekarski (2002) have shown that the design can be simplified to a simple random sample (srs) design with only a small increase in bias. The srs design is the most commonly used method, especially for state, regional, and national samples. The advantages of an srs versus a multistage design are simpler record keeping and analysis procedures and less cost.

The key to a list-assisted design is obtaining a list of all prefixes plus hundred banks that have at least one listed residential number. A *hundred bank* is the first five digits of a seven-digit phone number, excluding the area code. Saying this another way, it is the prefix plus the first two digits of the four digit suffix. The last two digits of a seven-digit phone number are the hundred bank because these digits consist of 100 possible numbers: 00 to 99. Once a list of five-digit numbers is obtained, a systematic or simple random sample of these numbers is selected and two random digits are added to make a seven-digit telephone number. How many numbers to select and the specifics of the selection process are discussed in Example 2 below.

A listing of prefixes and hundred banks with at least one residential number can be obtained from a number of commercial organizations. Haines and Co., Inc. has telephone directories for 75 major markets, including 9 states and the District of Columbia. The books give listed telephone numbers by street address or in numerical order. The latter directory is called a *telokey* and is exactly what is required to do a list-assisted sample. Another source is InfoUSA. They purchase telephone directories and other public lists, merge them, and sell the information for marketing and other purposes. You can purchase entire lists of seven-digit telephone numbers from them by area code, region, or other geographic divisions of interest.

One drawback of purchasing an entire list of telephone numbers and doing the sampling yourself, especially for a large population area, is cost. If the list is only to be used once and not amortized over a number of studies, it is usually cheaper to buy a list-assisted sample than an entire list. The two companies that most survey groups work with are Survey Sampling International and Marketing Systems Group/Genesys.

Example 2: A Directory-Based Community Telephone Sample

For a study that covers a small geographic area and is represented by one telephone directory, a convenient method is to select a directory-based

RDD sample. For this example, we will select numbers randomly from the telephone directory, drop the last two digits of the telephone number, and replace them with two randomly selected digits. Two assumptions underlie this method. One is that the telephone directory includes all existing telephone prefixes and that no new prefixes have been established since it was published. New prefixes could lead to potential bias because households that are assigned these numbers would not have a chance of being in the study. Let's assume the telephone directory is only a few months old, so this possibility is unlikely. However, when in doubt, we should call the telephone company and ask. A second assumption is that the unlisted telephone numbers are distributed proportionately among the listed numbers. If this assumption is not true and the unlisted numbers are clustered in a few prefixes, then the unlisted numbers would have a smaller probability of selection. Another concern is that in many geographic areas, prefixes are confined to specific geographic locations. If an area has a disproportionate percentage of unlisted numbers, it will be either overrepresented or underrepresented, which is likely to bias the final results. *We do not know of any studies that have tested this assumption; researchers either assume proportional distribution or ignore the potential problem.*

For this study, we will develop a systematic random sample, selected in stages. To do that, we need to estimate the total number of sample selections, compute a sampling interval, and select a sample starting point. This procedure is far simpler and more efficient than an srs. For example, let's assume the residential section of the Raleigh telephone directory is 440 pages long. Each page has four columns, and each column contains 104 lines. To select a simple random sample from this book, we would need to number every household telephone number on the 440 pages, a total of approximately 122,500; we would then select unique random numbers between 1 and 122,500. The hard part is the initial numbering. How long do you think it would take you to write the numbers between 1 and 122,500?[1]

Expression 8.1 is the formula for determining the number of sample selections needed for the proposed telephone study. Over the next few pages we are going to discuss Expression 8.1: (a) how to solve it, (b) the assumptions we are using, and (c) how we arrive at the numbers and percentages we will use to solve it. Because the expression, or parts of it, is used several times in this chapter, we discuss it in detail. In words, the expression means that to determine the number of sample selections, divide the desired number of completed interviews by the product of the percentage of the eligible telephone numbers that occur at each stage of sample selection and the percentage of residential numbers that become completed interviews.

$$
\begin{array}{c}
\textit{Number of} \\
\textit{sample} \\
\textit{selections}
\end{array} = \frac{\textit{Required number of completed interviews}}{\left(\begin{array}{c}\% \textit{ of residential} \\ \textit{telephone \#s in} \\ \textit{the phone book}\end{array}\right) * \left(\begin{array}{c}\% \textit{ of RDD} \\ \textit{telephone \#s that} \\ \textit{are households}\end{array}\right) * \left(\begin{array}{c}\% \textit{ of residential} \\ \textit{telephone \#s that become} \\ \textit{completed interviews}\end{array}\right)}
$$

(8.1)

The three components in the denominator correspond to points in the sample selection and data collection processes where ineligibles or noninterviews occur. The sampling rate must compensate for ineligibles and noninterviews.

Solving for the Number of Sample Selections

To determine the requisite number of sample selections, we need to think through the entire sample selection process and estimate how well the interviewers will do. First, we need to decide on the desired number of completed interviews. Assume we want 500. Second, we need to think through the sample selection process. We are going to select residential phone numbers from a hypothetical telephone directory. We will drop the last two digits and replace them with two random digits. One question we must ask is how good our sampling frame is. Is every listing a residence? Does every listing have the same chance of selection? To answer these questions we need to check the frame.

Exhibit 8.2 is a facsimile of part of a telephone directory. Look at column 2; not every line of type ends in a telephone number. Note, for example, that B. Edward Smith's listing takes up three lines: one line lists the name, the second lists a business address and telephone number, and the third lists the residence address and telephone number. We can see that our frame is not "pure"—there are some businesses and, although most people are listed on one line, some people's listings are longer. The more lines a person has, the higher the probability of selection or the larger the chance of being in the sample. To equalize the probabilities of selection and to estimate the number of ineligible lines, we need a decision rule. Our decision rule will be to select the phone number only if the sampled line contains a residential telephone number. For example, if the line in the third column that lists the names of C. George and Margaret Ann Smith fell into our sample, we would consider that line blank and treat it as ineligible. However, if we drew the following line, which lists their residential telephone number, we would take it and they would be selected in the sample.

We also need to estimate the proportion of business listings and blank lines. We do this by taking a sample from the directory. For this study 25 columns

Exhibit 8.2   A Facsimile of Part of a Telephone Directory Page

SMIT–SMITH	*AREA CODE 101*	*RESIDENTIAL SECTION*	*378*
SMIT Robert 231-0401	SMITH Brock 1775 Charles St. 271-1661	SMITH C Allen 272 Lane St 271-0408	SMITH D A 452 W Lane St. . . . 463-3629
SMITE A J 276-1112	B Charles 784 Falls Rd 221-7743	C Bruce 248 Brooks Ave 271-7003	D Billy & Sally 172 Lane St. . . 271-9103
	B Douglas High St Ashe 221-0909	C Dallas CPA 10 Main St 785-2838	D Carl 3334 Vernon 421-4746
SMITH—See Also Smythe	B Edward	Res RR 1 253-1180	Daryl D 4127 Shela Ln 483-1988
	Ofc 154 Flint St 826-1122	CF 7812-E Apple Blossom Ct. . . 482-6001	D Elvis 333 Drexel Ave 271-0369
A B 8403 Lawrence Rd 482-1364	Res 1225 Charles St 271-6665	C George & Margaret Ann	D Francis 6235 N Green St . . . 785-1111
A Curt 113 Adam Ln Ashe 221-0123	B Frank 621 Beaver Ln 463-7723	5720 Benton Rd 785-9902	D Glen 1534 Lakes Rd 482-4371
A Don 804 Monk Cir Mdvl 253-4444	B Greg & Michelle P	C Hal 421 Davis Rd 271-7801	D Hank 638 Lewis St 482-5300
A Ed 7821 Adam Ln Ashe 451-2233	6698 Prince Charles Rd 271-3434	C Ike 1013 Falls Lake Ashe . . . 221-2311	D Islam 3947C Pepper St. . . . 281-4400
A Floyd 831 Woodley Rd 785-4131	B Harry 686 Round St Mdvl 253-7246	C John 23 Wilber Hall 281-0028	D Justin 1221 Shellie Rd 482-2200
A Gilbert 7740 Lawrence Rd . . 482-9264	B Irv 9465-C Clarke Ln 785-5232	C Kurt Ofc 10 W Main St 271-6050	D Kirk 2604 W Green St. . . . 276-1662
A Helen 144 Wolfline Ashe . . . 451-7203	B John 81 Mullen St 281-8552	486 Brentwood Mdvl 253-6112	D Linda 767 Sherron St. . . . 483-1933
A Irene 1113 Winslow Ln 281-4455	B Kelly 4653 Bryant Cir 482-0013	C Luke 333 Weaver Ln 785-2331	D Mark 1517 Burn St 271-0839
A Jess 4455 Hillcrest Rd 785-9999	B Lawrence	C Molly 2517 Country Club . . . 463-4545	D Neil 516 Sanders Ct 271-5915
Anthony K 123 Lyons Rd . . . 223-7845	243-B6 Apple Blossom Ct 482-0065	C Nick 6442 Clarke Ln 785-6668	D Opra 946 Harvey St. Ashe . . 221-0009
A Paul & Wanda	B Ryan 781 Main St Ashe 221-6251	C Quinton 510 Adam Ln Ashe . . 221-9155	D Phillip 2854 N Clark 783-5815
322 Adam Ln Ashe 451-6547	B Steve 200 Newport Ct 787-7831	C Rick 954 Bryant Cir 482-6835	D Raymond 1735 Harris St . . 751-0435
A Robert 742 Smallwood Ln . . . 271-2801	B Thomas 154 Styles Rd 223-4654	C Scot 84 Prince Charles Rd . . 271-0003	D Yales 12 E Lane St. 251-6666

Note: Names, addresses, and phone numbers are fictional.

were randomly selected from the 440 pages and in each column, the blank lines and businesses counted. The total was 231, or an estimate of 8.9% of the lines.[2] We will use this information shortly.

Once we create a new telephone number by substituting digits, we must think through what can happen to this number. There are three basic possibilities: the number can be an eligible household, it can be ineligible (e.g., a business), or its status can be unknown. Let's talk about the last two possibilities first.

When we create a new telephone number, it may be a business telephone number, a disconnected number, or a nonworking number, all of which would be classified as ineligible. It is necessary to estimate the proportion of selections that will be ineligible when calculating the number of sample selections. The best way to do this is to consult the results of previous surveys. However, if this is the first time this procedure is being used in the area, the pretest results can be used as a guide.

Also affecting the number of sample selections is the estimated frequency of unknown status numbers, which ring when called but which no one ever answers. Many researchers have interviewers let a number ring at least seven times before hanging up. However, after repeated calls at different times of the day and on different days of the week, some numbers may still result in a "ring-no-answer." Groves and Kahn (1979) report making 12 contact attempts at this type of number, and they conclude that very few of these telephone numbers were eligible households.

In addition, not all eligible households can be interviewed. A number of things, such as refusals, or language problems, may preclude an interview. Or the selected respondent may be ill, out of town, or unavailable for other reasons. Nevertheless, the response rate, or the number of completed interviews from the total number of eligible households, must be estimated. Again, this is best done from previous surveys or from pretest results. (Estimates based on pretests may have high sampling variances. Why is this true?)

We now return to Expression 8.1 to determine the number of sample selections required to yield 500 completed interviews. We use the results of the sample of columns as the estimate of the first term in the denominator $(1.00 - .089 = .91)$. In addition, let's assume we found that 28% of the RDD numbers in our pretest were ineligible and that our response rate was 70%. Putting these numbers in the expression gives the following result:

$$\text{Number of sample selections} = \frac{500}{.91 \ * \ .72 \ * \ .70} = \frac{500}{.45864} = 1090$$

This means that we must select 1,090 lines from our hypothetical telephone directory to end up with enough telephone numbers to achieve 500 completed interviews. If our assumptions are correct, the results will be as follows:

Of the	1,090	telephone directory line selections,
	98 (9%)	will be ineligible (i.e., blank lines or businesses), giving us
	992 (91%)	residential telephone numbers.

We convert these 992 numbers into new telephone numbers by dropping the last two digits and replacing them with two random digits.

Of the	992	telephone numbers, we estimate that
	278 (28%)	will be ineligible (i.e., businesses, disconnected, and nonworking numbers), and that
	714 (72%)	will be eligible households.
Of the	714	eligible households, we further assume that
	214 (30%)	will result in noninterviews (i.e., refusals, unknown status, noncontacts, etc.) and that
	500 (70%)	will be interviews.

Selecting the Sample

Now let's focus on making the telephone directory selections. The hypothetical directory has 440 pages with four columns per page and 104 lines per column. We want to select a systematic random sample. The first thing to notice is that we need to make more selections than there are pages. In fact, we need to make more than two but less than three selections per page. How do we do this? Note that there are four columns per page, or a total of 1,760 columns. Let's see how the total number of selections compares to the total number of columns: 1,090/1,760 = .619. According to this calculation, we should take one selection from 62% of the columns or one selection from every 6 of 10 columns. These are not easy intervals to work with. We want to make the selection process easier but still random. Notice that 62% is close to 66.7% or 2/3. One possibility is to take more selections than we need and then to randomly delete some of the selections. The remaining sample is still a random sample because a random sample of a random sample is a random sample.

The easiest way to select 2 of every 3 columns is probably to delete 1 of every 3 columns. We do this by selecting a random number between one and three. Then, we add 3 to this number until we exceed the number 1,760. Each of the columns selected in this manner is excluded from sample selection. Depending on the random number selected, this process would delete approximately 587 columns, leaving us 1,173 columns from which to make a selection. To select from each of the 1,173 columns, we select a random number between 1 and 104 (number of lines in each column). Using a table of random numbers, we select the number 017. In each selected column, then, we examine the seventeenth line, using a template that measures the distance from the solid line at the top of the column to the seventeenth line. If it is a residential household, it falls into our sample; if it is a business or a blank line, we consider it ineligible and go on to the next selected column.

Refer to Exhibit 8.2 again. Assume that column 3 is deleted from the sample. Thus, we examine the seventeenth line in columns 1, 2, and 4. The seventeenth line in column 1 lists the residential telephone number of A. Paul and Wanda Smith, who live in Ashe. We select their telephone number, B. Steve Smith's telephone number from the second column, and D. Raymond Smith's phone number from column 4. For all the phone numbers selected, we drop the last two digits and substitute two random digits. From a table of random numbers we have selected the digits 07, 79, and 91. Replacing the last two digits of our selected phone numbers with these random digits yields the following new telephone numbers: 451-6507, 787-7879, and 751-0491. This is the procedure that we follow to select all the remaining numbers.

If our sample estimates are correct, we expect to encounter about 106 blank lines and businesses and 1,067 residential telephone numbers. If our other assumptions are correct, we can expect 28% of the new telephone numbers to be ineligible ($n = 299$) and we would end up with approximately 768 eligible households. If our survey **response rate** is 70%, we would complete 538 interviews. If a completed interview costs about $30, these 38 additional interviews would add $1,140 to our survey costs. Not a trivial amount. Thus, our goal is to manage the sample so that the number of final interviews is close to 500. How to achieve this goal is illustrated in the discussion of Exhibit 8.3.

It is extremely important to realize that the proportions we inserted in the denominator of Expression 8.1 are only *estimates* of what we expect will happen. As we go through the process of selecting numbers and attempting interviews, the outcome at any stage can be better or worse than we estimated. We could end up with fewer ineligible numbers and a better response rate than we expected. This would give us more sample numbers than we need. We could also get more ineligibles than we counted on and a poorer response rate, which would give us too few sample cases to achieve 500 interviews. Because any number of combinations are possible outcomes, it is important for the researcher to keep close watch on the process as it unfolds and to take corrective actions when necessary.

The following example illustrates how we can best achieve the desired number of completed interviews through the process of subsampling and monitoring. This process is illustrated in Exhibit 8.3.

Assume that our estimates of the proportion of blank lines and businesses are correct and the sample selection yields 1,067 RDD numbers. Each telephone number released to an interviewer must be thoroughly worked to maintain our random sample and minimize bias. Because our particular sample of numbers could yield fewer ineligibles and/or a higher response rate, we do not want to release all 1,067 numbers at once. We want to subsample the numbers, see what happens to those cases, and use that information to plan activities for the next week or few weeks. To do

Exhibit 8.3   Management of RDD Sample Numbers

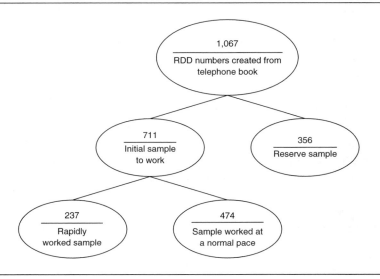

this, we try to anticipate the best possible outcome for our numbers. We can think of this as a sample distribution problem; that is, try to determine what outcomes might be three standard deviations on the positive side from our 72% estimate of eligible numbers and our 70% estimate of a response rate. Because we do not know this information, we need to make an educated guess. Let's use 85% for eligible numbers and 80% for a response rate. We can again use Expression 8.1, without the proportion for blank lines and businesses, because that phase has been completed. The result is:

$$Number\ of\ sample\ selections = \frac{500}{.85 * .80} = 735$$

We want to select a systematic random sample of 735 numbers from our sample of 1,067 numbers. To do this, we go through the same mechanics as we did earlier in the selection of columns from the telephone directory. We find that 735/1,067 = .689, which is very close to .667 or 2/3. Even though .667 × 1,067 = 711, or 24 fewer numbers, this is a reasonable number of cases to begin with. After selecting the 711 numbers, it would be a good strategy to split them into two samples. One sample would be worked thoroughly at a rapid pace to monitor our assumptions and to determine if additional sample cases are needed. The other sample would be worked concurrently, but at a more normal pace. The rapid sample might be a one-third subsample of the 711 cases, or 237 telephone numbers. We would want to contact each case a minimum of seven times over a 5- to 7-day period. This random subsample should give us a good indication of the number of ineligibles and interviews we can expect from the initial release of 711 numbers. As the data collection period progresses, we continue to monitor the results against our assumptions. In reserve, we have a random subsample of 356 numbers (1,067 − 711 = 356). Random subsamples can be taken from this pool and released as required, or, if necessary, all numbers can be released at once.

Example 3: Other RDD Telephone Samples

There are a number of other ways to select an RDD sample. One method is to determine the telephone prefixes that cover the area and add four random digits. This makes the sampling tasks easier but the interviewing is more difficult than in the first two examples, as we will see shortly.

Exhibit 8.4   Prefix Codes for Raleigh and Selected Nearby Cities

Location	Prefix Codes
Apex	303, 362, 387
Cary	319, 380, 460, 467, 469, 481, 677
Garner	662, 772, 779
Knightdale	217, 266
Morrisville	467, 469, 481
Raleigh	212, 231, 233, 250, 301, 302, 420, 501, 505, 508, 512, 515, 516, 518, 546, 571, 662, 664, 676, 713, 715, 733, 737, 740, 755, 772, 779, 781, 782, 783, 787, 790, 801, 821, 828, 829, 831, 832, 833, 834, 836, 839, 840, 846, 847, 848, 850, 851, 856, 859, 860, 870, 872, 876, 878, 880, 881, 890, 899, 954, 971, 976, 981
Wendell	266, 365
Zebulon	269, 365

The front section of a telephone directory typically lists the residential telephone prefixes in the local calling area. Exhibit 8.4 lists the prefixes for Raleigh and selected nearby cities for the one annual telephone directory that we defined as the area of earlier interest. It is worth noting that the number of prefixes correlates with the size of the city. Raleigh, the largest city, has the greatest number of prefixes, followed by Cary and then Garner. Also, many, but not all, cities have unique prefixes. Note that Apex has its own set of prefixes but Morrisville has a subset of Cary's prefixes; Knightdale, Wendell, and Zebulon have some overlap; and Garner has a subset of the Raleigh prefixes. Unique prefixes allow the researcher to focus on a specific area if the researcher so desires. If the prefixes overlap and the researcher is interested in only one city, screening questions about geographic residence are required at the start of the interview to eliminate the ineligible respondents.

There are 77 unique prefixes in the geographic area. To develop the RDD telephone numbers, we simply add four random digits to the prefixes. But how many numbers do we need? We can use some of the assumptions in Expression 8.1. If we want 500 interviews, the numerator is the same. Because we are not selecting numbers from a telephone directory, we don't need to deal with blank lines and businesses. We also omit the percentage of residential numbers in the telephone directory because that does not apply here.

We do need to estimate the number of ineligible numbers and the response rate. We can assume a response rate of 70%, as we did earlier, so that leaves one unknown. Can we use 72% again as the proportion of eligible numbers? No, because we are using a much different procedure to generate telephone numbers. The major difference is that we are adding four random digits to a prefix rather than dropping two digits from an existing telephone number. This may not seem, on the surface, like much of a difference, but we will show shortly that it can be.

The four digits after the prefix are called a suffix. Each prefix can have 10,000 potential suffixes attached to it—from 0000 to 9999. When we select a phone number from the telephone directory and drop the last two digits, we know that the telephone company has assigned numbers using the first two digits of that suffix. Earlier, we selected the telephone number 787-7831. We know that for prefix 787 the telephone company is using the suffixes 7___ and 78__ . The latter is a hundred bank. When we add four digits to a prefix, we don't usually know the suffixes that the telephone company is using until we call the numbers. Thus, using four-digit random suffixes leads to more nonworking telephone numbers.

We can estimate the proportion of ineligible telephone numbers when we add four random digits to a prefix by using census data and knowing the number of unique prefixes. Let's assume the Raleigh geographic area of interest had 144,621 occupied households 3 years prior to doing this study. We check with census data and find that 96.5%, or 139,559, of these households had telephones. Because our census data is about 3 years old, we need to estimate household increases in the area. One way of doing this is to look at growth in the previous 10 years and assume that the area is growing at the same rate. Another way is to get estimates of growth from real estate firms, banks, or government agencies. By consulting census data we find that in the previous 10-year period, the urban population of Wake County increased by 49% and the rural component by 19%. Using this past decade as a guide, we assume there has been about 14% growth, which translates to 159,098 households with telephones. Exhibit 8.4 shows 77 unique prefixes in the area. Because each prefix can have 10,000 unique suffixes, there can be a total of $77 \times 10,000 = 770,000$ unique telephone numbers in our geographic area. The estimated number of eligible telephone numbers is 159,098/770,000 = .2066. The number of four-digit random numbers we need to create is determined by using our new assumptions in Expression 8.1.[3]

$$\begin{array}{c} \textit{Number of} \\ \textit{sample} \\ \textit{selections} \end{array} = \frac{500}{.2066 * .70} = \frac{500}{.1446} = 3{,}457$$

This result, much larger than the 992 numbers we needed from the directory-based method used in Example 1, indicates that we need to call approximately seven numbers to get one completed interview. Many of the calls and much of the budget will be spent in eliminating nonworking numbers.

Continuing with our example, we have 77 prefixes and we estimate that we need 3,457 numbers to get 500 completed interviews. The number of four-digit random numbers we need to create for each prefix is 3,457/77 = 44.89 or 44.9. Rounding up to 45 changes the total by 8, an insignificant difference: 45 × 77 = 3,465. Thus, we select 45 unique four-digit random numbers for each prefix. We select the digits from a table of random numbers or generate four-digit random numbers by computer. Because our total of 3,465 is an estimate, we would create all the telephone numbers, take a systematic random subsample of the numbers, and call and monitor the subsample results as we did in Example 2 and illustrated in Exhibit 8.3.

In this example, interviewers will spend a considerable amount of time eliminating nonworking numbers. However, there are aids to help cut down on the number of nonworking numbers. Haines & Co. (Middleburg Heights, OH), publishes criss-cross telephone directories for selected cities. Rather than listing patrons alphabetically, the directory lists them by street address and in ascending numerical order by prefix and suffix. This latter listing is helpful in finding suffixes that are not being used and can be eliminated in the selection process. For example, assume the two smallest prefix numbers are 212 and 217. Because the suffixes are listed in ascending order, we may note that the last listed telephone number in prefix 212 is 212-4356 and that the next listed telephone number is 217-0018. We might, then, conclude that in prefix 212 suffixes 5_ _ _, 6_ _ _, 7_ _ _, 8_ _ _, and 9_ _ _ are not being used. If this is correct, we can exclude 5,000 nonworking numbers from our potential list of telephone numbers. We would make similar checks through our entire list.

In areas where criss-cross directories are not available, sometimes it is possible to scan a large sample of the telephone directory pages to create a pseudolisting in ascending order. This should be done cautiously, however, because when a suffix is excluded, respondents with those telephone numbers have no chance of selection. Finally, large blocks of numbers can be eliminated by checking to see whether the large employers in the area have been assigned their own prefix. In Raleigh, for example, North Carolina State University uses prefix 515 and 513, and the State of North Carolina uses prefixes 715 and 733. We could safely eliminate these 40,000 potential telephone numbers. Some nonworking and business numbers will remain after all of these processes. The business numbers are best identified and eliminated by making the first contact attempt during the day, Monday through Friday.

National RDD Sample

Beginning in the 1990s, telephone systems worldwide and in the United States underwent major changes. New technologies have been a major impetus and the transformations are still occurring as we begin the 21st century. What systems will look like in 2010 is anyone's guess. Some major changes that have occurred in the U.S. system are the number, distribution, and density of residential telephone numbers by hundred banks. Tucker et al. (2002) report that the number of hundred banks with one listed residential number in 1990 was 38%, in 1999 it had decreased to 30%. In 1990, 62% of the 4.35 million hundred banks had no listed residential numbers; in 1999, 70% of the 7.72 million hundred banks had no listed residential numbers. In terms of density, in 1990, 14.9% of all hundred banks had 50 or more listed residential numbers, in 1999, it was 3.5%. These changes affect the efficiency of many designs.

Many of the procedures discussed in the previous sections can be used to select a national RDD telephone sample. The differences are the use of a national sample frame and the selection of numbers for some designs in a two-step process. It would be possible, but not very efficient, to select a national sample from telephone directories. There are about 5,100 telephone directories published annually in the United States. These books come in different sizes and they are published at different times of the year. Obtaining copies and selecting a sample from these directories would be a very expensive and time-consuming task. Fortunately, there are better methods.

In the past decade, list-assisted srs methods were the design of choice. The design and selection procedures were as we outlined in Examples 1 and 2. Another method that was used extensively prior to list-assisted methods but which is slightly less efficient, is a method proposed by Mitofsky and Waksberg (Waksberg, 1978). This method uses a data tape produced by Telcordia Technologies (TT). TT produces an updated listing quarterly, called NPA/NXX Active Code List, of every telephone area code and working prefix in the United States. In mid-2003, there were 271 area codes and 86,893 area code and prefix combinations. We could use the TT data tape as the sampling frame for a national RDD survey. Selecting a sample of area codes and prefix combinations and then adding a four-digit random suffix, however, suffers from the same problem encountered in Example 3: many nonworking numbers. Using this approach, Tucker et al. (2002) found that less than 15% of the sampled numbers were working residential numbers. To overcome this problem, Mitofsky and Waksberg (Waksberg, 1978) proposed a method that increases calling efficiency but is still not quite as good as an srs list-assisted design. The sample selection and interviewing are done in two stages. Initially, the researcher must decide the total number of interviews required, how many of these will be done with first-stage sampling units, and how many will be done with second-stage sampling units. Let's work through an example because it gives an indication of how to do a multistage sample and it illustrates the principles of probability proportional to (some measure of) size sampling.

Assume we want 1,600 completed interviews in 400 geographic areas. (We define geographic area shortly.) We create telephone numbers for first-stage interviews, conducted in all 400 geographic areas, by taking a systematic random sample of area codes and prefix combinations from the TT data tape. How many of these area code–prefix combinations do we select? To determine this, we divide 400 by the proportion of eligible numbers found by Tucker et al. (2002). The result is

$$Number\ of\ first\text{-}stage\ sample\ selections = \frac{400}{.15} = 2,667$$

Thus, we need to select a systematic random sample of 2,667 area code–prefix combinations from the TT total of 86,893.[4] To the six-digit area code–prefix combinations, we add four-digit random numbers to create the telephone numbers. The 2,667 numbers should yield about 400 residential telephone numbers. However, we know that not all of these will be completed interviews. Having estimated a 70% response rate, we can expect about 280 interviews.

The sample is split into two stages because we know from our previous examples that once we find a residential telephone number, there are likely to be more such numbers in that area code and prefix combination. The second stage of the Mitofsky-Waksberg design is to keep the first eight digits of the numbers that are determined to be residences in the first stage, that is, the area code, the prefix, and the first two digits of the four-digit suffix. To this eight-digit number, we add two-digit random numbers until we contact a fixed number of additional residential telephone numbers using the eight-digit number. For example, assume that the number 919-387-7892 yields a residential telephone number. To create the second-stage telephone numbers, we drop the last two digits and keep substituting two other random digits until we identify a fixed number of additional residential telephone numbers.

To complete our example, how many other numbers must we create? In the first stage, the 2,667 numbers yielded 400 households and resulted in 280 interviews. We need to complete (1,600 − 280 =) 1,320 more interviews using the 400 first-stage telephone numbers. This means an average of (1,320/400 =) 3.3 interviews for each residential number identified in the first stage.[5] How many second-stage numbers do we select? Tucker et al. (2002) found that 49% of the numbers at the second stage were households. If we use a 70% response rate, Expression 8.1 can be solved as follows:

$$Average\ number\ of\ second\text{-}stage\ sample\ selections = \frac{3.3}{(.49 \times .70)} = 9.62$$

Creating an average of 9.62 telephone numbers per first-stage household number would result in calling approximately 3,848 (9.62 × 400) more numbers to obtain 1,320 interviews. The second-stage numbers are created by dropping the last two digits of the residential numbers identified in stage 1 and substituting groups of unique two-digit random numbers until the fixed number of residences is identified.

The Mitofsky-Waksberg design is a two-stage cluster sample with the probabilities of selection being equal after the second stage. The probability of selection for a group of 100 numbers at the first stage of selection is proportional to the number of residential numbers in the hundred bank. Thus, the more residential numbers in the hundred bank, the higher the probability of selection. For this reason, a fixed number of residential units must be selected in the second stage to equalize the overall probabilities of selection.

We can illustrate this as follows. Assume telephone number A is in a group of 100 numbers that has 80 residential telephone numbers and telephone number B is in a group of 20 residential numbers. If one telephone number is selected from each group of 100 numbers, the probability of Group A being in the sample is 80/100 = .80 and for Group B it is 20/100 = .20. At this first stage of selection, the likelihood of Group A being in the sample is four times greater than that for Group B. At the second stage of selection, we want to identify four more residential numbers so that we call a total of five residential numbers in each group. The overall probability of selection of a number in each group is the product of these two probabilities. As we illustrate below, the probabilities are equal after stage 2.[6]

$$Group\ A = \frac{80}{100} * \frac{5}{80} = .8 * .0625 = .05$$

$$Group\ B = \frac{20}{100} * \frac{5}{20} = .2 * .2500 = .05$$

This type of design introduces a number of complexities that are not part of simple random samples. First, the formula used to compute the sample variances must take the two-stage design into consideration (Kalton, 1983).

Second, because people who live in the same area tend to be more similar in attitudes and behaviors than the population as a whole, this type of sample has cluster effects because people within a cluster tend to be more homogeneous than a random sample of the population, and it will underestimate the population variance. This results in a **design effect**—the ratio of the variance of a cluster sample to the variance of a simple random sample of the same size. A third complexity is that the sample must be worked sequentially and a great deal of monitoring is required to ensure that the fixed number of residential households is not exceeded within the clusters (hundred banks). Finally, in some cases the fixed number of residential numbers is not achieved for a cluster. Strictly speaking, when that happens, a weight needs to be applied that is the ratio of the target number of residential phone number to the actual number identified. Why then is the Mitofsky-Waksberg procedure used? Multistage cluster samples are the preferred choice when the design effects are small for the key variables and when sampling and data collection costs are significantly lower for a cluster sample than for a simple random sample.

Selecting Respondents Within Households

A sample of telephone numbers is a sample of households. It is possible to collect fairly accurate information at the household level on activities such as the number of doctor visits by all household members in the past 2 weeks; the number of overnight hospital stays in the past 12 months; the number and types of durable goods purchased in the past 6 months; the number of automobiles owned and total miles driven in the past 12 months. In conducting such a survey, we call the telephone number and we tell the person who answers what the survey is about. We ask to speak to the person in the household who is most knowledgeable about these activities. This type of study seeks to interview the adult who can provide accurate information.

Many times, however, we want to generalize our results to, or estimate characteristics for, a population of individuals. If we were to interview the person who answers the telephone, we would end up with a disproportionate number of teenagers, women, people who are active in community affairs, and people who run a business from their home. To satisfy the ability to generalize to or estimate characteristics of individuals, we usually select randomly one individual per household and weight the interview data by the number of eligible respondents in the household. The weighting is required because households are clusters of eligible individuals. When we select one individual per household, people in households with two or more eligible respondents don't have the same probability of selection as do people in one-person households. The person in a one-person household has a 1.0 or 100% chance of selection because that individual is the only eligible respondent. People in a two-person household each have a .5 chance of selection; people in a three-person household each have a .33 chance of selection, and so on.

The classic within-household selection procedure was designed by Kish (1965) for face-to-face surveys and has been adapted for telephone surveys. In telephone surveys, after a brief introduction, the interviewer asks the respondent to list all the eligible household members by gender and age. The interviewer must then number each eligible household member starting with the males in descending order by age and then the females in the same manner. Using a table of preselected random numbers—in a two-person household the number is a 1 or 2—the interviewer selects the individual whose number corresponds to the preselected random number. While the Kish method is unbiased, it requires a significant amount of time at the beginning of the interview. Many respondents are suspicious of detailed household enumeration questions before survey legitimacy or rapport has been established. For that reason, many investigators believe this selection procedure leads to higher refusal rates.

Other procedures were proposed by Troldahl and Carter (1964) and Bryant (1975) in an attempt to make household enumeration less invasive. In these procedures, the adult who answers the telephone is asked two questions:

1. How many persons 18 years or older live in your household, counting yourself?

2. How many of them are men (women)?

One of four or seven selection matrices is randomly assigned to each sampled telephone number.[7] The selection matrices do not require a listing of adults. Instead, the interviewer asks to speak with the "oldest" or "youngest" man or woman, or simply a man or woman, depending on the household composition. In an experimental comparison of these methods, we did not find the Troldahl-Carter-Bryant methods to be better than the Kish method (Czaja, Blair, & Sebestik, 1982). Our preference was the Kish procedure.

A more recent technique was proposed by O'Rourke and Blair (1983) and Salmon and Nichols (1983). These procedures do not require household enumeration. Interviewers simply ask to speak to the eligible member of the household who had the "last" or will have "the next" birthday. The last-birthday methods works as follows:

```
Hello,  I'm  _____  calling  from  the  University  of
Maryland.  We are doing a study to find out how people in the
Washington  Metropolitan  Area  feel  about  things  that  may  con-
cern  them.  Your  household  was  chosen  at  random  to  partici-
pate.  For  this  study,  I  need  to  speak  with  the  adult  living
in  your  household,  who  is  18  or  older,  and  had  the  most
recent  birthday.  Who  would  that  be?
```

IF THE PERSON DOES NOT KNOW ALL THE BIRTHDAYS, ASK:

```
Of  the  ones  you  do  know,  who  had  the  most  recent  birthday?
I  need  to  speak  with  that  person,  please.
```

Both methods seem to give each eligible member an equal chance of selection. The last-birthday method probably has less selection error than the next-birthday method because it may be easier to recall a past event than to know an upcoming event, unless the future event is much closer in time. Methodological studies are required to determine if this hypothesis is true and to better evaluate the reliability and validity of these selection procedures. Oldendick, Bishop, Sorenson, and Tuchfarber (1988) did a methodological study comparing the Kish and the last-birthday method. The authors compared refusal rates, demographic characteristics of the samples, and substantive responses to questions. They found very few statistically significant differences, and they concluded that the results did not differ by method. More such studies need to be conducted.

Recently, a method was proposed by Rizzo, Brick, and Park (2004) which takes advantage of the fact that approximately 85% of U.S. households have only one or two adults. When a household is contacted, the first question is: How many adults age 18 or older live in this household? In those cases where the answer is one, that person, obviously, is the selected respondent. When there are two adults, half the time take the person on the phone and half the time ask for the other adult resident. In only approximately 15% of the households is a more complicated procedure needed. In those cases, either "Next Birthday," or one of the other methods described can be used.

Example 4: A List Sample of Students

Studies of university students are frequently conducted by professors or by other students. We already discussed one method for conducting such studies, a telephone survey.[8] For this example, however, let's assume that we have very little funding and time is not a major factor. Therefore, we have decided to do a mail survey using campus mail as much as possible. (In Example 5, we discuss another possibility—selecting a sample of classes for a group-proctored survey.)

Selecting a Frame

The registrar's office has the most current information on enrolled students. We have found these offices to be cooperative when the survey has a serious academic focus and the collection of information can be useful to other campus groups. Almost all universities and colleges keep student information in computer files; these files are usually in alphabetical order. Let's assume we want to study undergraduates only and that our university has 20,000 undergraduate students: 8,000 freshmen, 6,000 sophomores, 3,000 juniors, and 3,000 seniors. We want to select a sample of 1,000. One method would be to ask the registrar's office to select a simple random sample of 1,000 undergraduate students from its computer file. With a sample of this size, we should get approximately 400 freshmen, 300 sophomores, 150 juniors, and 150 seniors, although there is no guarantee that the sample will be proportionate to the population by class year.

Deciding on a Sampling Method

There is a better way to select the sample if key characteristics of sample members (e.g., year in school, gender) are likely to be related to important dependent variables. It is called **stratified sampling**. Sudman (1976) presents four reasons for stratifying a sample:

1. Groups are of interest for purposes of analysis

2. Variances differ by group

3. Costs to conduct interviews vary by group

4. Amount of prior information differs by group

In stratified sampling, we want to arrange the population into groups of similar individuals. We want the elements within groups to be homogeneous and the groups to be different from other groups, or heterogeneous. The grouping variable or variables must be related to the dependent variable, otherwise the effort involved in stratifying the list will be a waste of time.

Implied stratification is one method of improving sampling efficiency by ensuring proportional distribution of the sample on specified variables.

Returning to our example, let's assume that we believe the results for our key dependent variables will differ by year in school. Therefore, instead of selecting a simple random sample from a list arranged alphabetically, we would ask the registrar's office to rearrange the list by year in school. The first 8,000 students would be the freshmen, followed by the 6,000 sophomores, who would be followed by the juniors, and then by the seniors. We then select a systematic random sample from the list. Because the population size is 20,000 and the sample size is 1,000, our sampling interval is 20. We select a random start between 1 and 20, begin with that number, and keep adding 20 to each selected number until we exhaust the list. This gives us a sample that is proportional to the population size. By using implied stratification in this way, we are assured that the sample will include about 400 freshmen, 300 sophomores, 150 juniors, and 150 seniors.

We could have the list sorted on multiple variables. For example, if we believe gender is also an important variable, we could arrange the list by year in school and by gender: the male freshmen might come first, then the female freshmen, followed by the male sophomores, then the female sophomores, and so on. Selecting a systematic random sample would assure proportional representation by gender and year in school.

We explained in Chapter 7 that the optimum design for comparing groups is equal sample sizes in each group. In this example, the sample sizes by year are unequal. To select equal numbers of students within each year in school, we would divide the students into four groups representing the four school years. Within each group, we would select a systematic random sample of 250 based on a sampling interval of 1:32 for freshmen, 1:24 for sophomores, and 1:12 for juniors and seniors. This is a *disproportionate stratified sample*. While the sample sizes and the number of completed interviews may be equal for each group, the sampling ratios and probabilities of selection for students by year in school are different. When we analyze our data, we can report the results by year in school, for example the proportion of freshmen, sophomores, and so forth, that had more than three alcoholic drinks last week, but we cannot simply combine data from the four groups because the probabilities of selection are different. To do that, we must weight the data. The equal probability of selection sampling ratio for all students is 1:20. Because freshmen and sophomores are undersampled, they must be given a weight greater than 1, and because juniors and seniors are oversampled, they must be given a weight less than 1. Weighting the data in this manner would allow us to combine the information from all groups to report an estimate of the proportion of all undergraduates who had more than three alcoholic drinks last week.

Example 5: A Sample of University Classes

A very efficient method of sampling students is to select a sample of classes. This method takes advantage of natural clusters and captive audiences. If the average class size is 24, then in the span of time that it takes to complete 1 questionnaire, we can obtain 23 more. The major obstacle to this method is obtaining the cooperation of the professors for the selected classes.

Defining the Population and Preparing the Frame

The process begins, as usual, by defining the population—for example, undergraduates—and then obtaining a suitable frame, such as a list of current classes. We need to ensure that classes that may be excluded from the list are included in the frame and that classes which have been canceled are excluded. Also, we need to eliminate duplication. Classes that have both a general lecture and affiliated labs or discussion sections need to be identified. If the lecture contains many students, say more than 100, and the labs are closer to the average class size, we want to keep the lab sections in the frame and exclude the large general lectures. The reason is that the large lectures will have large intraclass correlations that will increase the estimates of variance.

The schedule of courses at North Carolina State University lists classes alphabetically by department and, within departments, lists courses in ascending order from introductory courses to the most advanced. We will select classes with equal probabilities using a systematic random sample. An alphabetic listing provides no benefits for stratification. If we believe that attitudes or behaviors may differ by college, we would ask the registrar's office to run a new frame that lists the classes by college and, within each college, in ascending order from lower- to upper-level courses.

Solving for the Number of Sample Selections

Before selecting the classes, we need to determine the average number of students in the classes on the frame, the average number of classes taken by undergraduates, the anticipated cooperation rate of professors and students, and, finally, the required number of completed interviews. The average class size and the average number of classes taken can probably be obtained from the registrar's office. Only rough estimates are required. The most important piece of information is the average number of students per class, which, if unknown, can be estimated by multiplying the number of enrolled students by the average number of classes taken and dividing by the number of classes offered. This yields:

$$Average\ class\ size = \frac{20{,}000 * 4}{3{,}320} = 24.1$$

Assume we want 1,200 completed interviews. If the average class size is about 24, we need to obtain interviews in 50 classes. We know by now, that not every professor approached will agree to let us administer the questionnaire in his or her class. Thus, we need to estimate a cooperation rate. Again we look to similar past surveys or to the pretest. In a similar survey conducted at the University of Maryland, College Park Campus, 65% of the professors cooperated, as did 99% of the students in their selected classes. If we use a 65% cooperation rate, we would need to select 77 classes using systematic random sampling.

Maintaining the Design Integrity

While this design is very efficient and economical, the researcher must be aware of three additional considerations to maintain the integrity of the design. One, this procedure results in a probability sample of all enrolled students. However, the student probabilities of selection are not equal because students enrolled in more classes have a higher chance of selection than those enrolled in fewer. For this reason, proper analysis of the data requires that each student receive a weight. The weights can be adjusted relative to each other so that the total weighted sample size equals the total unweighted sample size. Because the average number of classes taken is four, we give those students who are enrolled in four classes a weight of 1. Those enrolled in more than four classes would be given a weight of less than 1, and those who are enrolled in fewer than four classes would be given a weight of greater than 1. The information about the number of classes each student is enrolled in has to be asked in the questionnaire. Omitting this question will leave you in a position of not having the information you need to properly weight the data.

Two, as we all know, all students do not attend every class, for a multitude of reasons. If there are differences between those present and those who are absent on the dependent variables, it will be important to include the nonattenders in the survey estimates or else the final results may be biased. Following up to obtain interviews from nonattenders is much more expensive than administering and collecting questionnaires in a classroom. The researcher will want to follow up a sample of nonattenders by mail or telephone, depending on the time schedule or budget. Neyman (1934) devised a method in which the researcher calculates the cost of a followup interview and the cost of a classroom interview, and the square root of the ratio of these costs determines the percentage of nonattenders who need to be followed up. In the analysis of the data, the subsample of nonattenders are weighted to reflect all nonattenders and, thus, the final sample estimates are unbiased.

Three, a random sample of clusters is not a random sample of elements. In a cluster sample, we would like the clusters to be made up of heterogeneous elements so that a sample of clusters would reflect a sample of the population. However, natural clusters typically reflect similarities in the population rather than diversity. In effect, students enrolled in introductory sociology are more likely to be freshmen and sophomores than to reflect a random sample of all enrolled students. The extent to which clusters are more similar in composition than a random sample of the population is what is referred to as the design effect of a cluster sample. This design effect causes a cluster sample to be less precise than a srs of the same size.[9] The extent to which it is less efficient is determined by the ratio of the variance of the cluster sample divided by the variance of an srs of the same size. This calculation is beyond the scope of our discussion and interested students should consult Kalton (1983).

Notes

1. The residential listings of some telephone directories are available for a fee as computer files from commercial organizations. Using a computer file would make it easy to select a simple random sample.

2. In Chapter 7, we explained that the larger the sample size, the more confidence we can have in the sample estimate, all other things being equal. A sample of 25 is very small and has a large sampling error. The reader must keep in mind that when using small samples for planning purposes, the final results may differ significantly from the planning sample.

3. The terms for the percentage of blank lines and businesses are omitted from the expression because they are not applicable to this method of creating RDD numbers.

4. This was the number of combinations in October 2003. Interestingly, the number of combinations in December 1993 was 44,129.

5. We use 400 because it is the number of households identified in stage 1. This is the important number, not the number of completed interviews.

6. This example has been simplified and these are not, technically, the overall probabilities of selection. The sampling fraction at the first stage of selection must be multiplied by the probability .05. The final overall probability of selection, however, will be the same.

7. Selection matrices 5, 6, and 7 give higher probabilities of selection to males because males have lower cooperation rates than females.

8. A possible sampling frame would be the published university directory of students. A sample of names could be selected in the same manner in which we selected our directory-based RDD sample. In this example, however, we would not want to create random-digit telephone numbers.

9. The confidence intervals or sample standard errors are larger for cluster samples than for srs samples of the same size. Thus, the estimate of the population value for a variable is less precise.

9

Reducing Sources of Error in Data Collection

This chapter considers how weaknesses in different aspects of data collection can affect the accuracy of our results. There are two main sections: concepts of error and methods to reduce error. The first section introduces the general idea of survey error, discusses the concepts of bias and variance, and provides an overview of unit and item nonresponse. The main measures of survey quality that these core concepts suggest are then summarized. We note the increase in unit nonresponse in recent years and suggest a framework for selecting procedures to address that problem.

This background sets the stage for the second part of the chapter in which we consider methods to reduce the sources of error.[1] Some sources of error are common to both interviewer-administered and self-administered surveys. We will deal with interviewer-administered surveys first, and note issues common to both data collection modes. Last, we take up issues peculiar to self-administered modes, such as conventional mail or Internet data collection.

The Origins of Error

Imagine the perfect sample survey. The survey design and questionnaire satisfy all the research goals. A sampling frame is available that includes accurate information about every population member. The selected sample precisely mirrors all facets of the population and its myriad subgroups. Each question in the instrument is absolutely clear and captures the dimension of interest exactly. Every person selected for the sample is contacted and immediately agrees to participate in the study. The interviewers conduct the interview flawlessly, and never—by their behavior or even their mere presence—affect respondents' answers. The respondents understand every question exactly as the researcher intended, know all the requested information, and always answer truthfully and completely. Their responses are faithfully recorded and entered, without error, into a computer file. The resulting data set is a model of validity and reliability.

Except for trivial examples, we cannot find such a paragon. Each step in conducting a survey has the potential to move us away from this ideal, sometimes a little, sometimes a great deal. Just as all the processes and players in our survey can contribute to obtaining accurate information about the target population, so can each reduce that accuracy. We speak of these potential reductions in accuracy as sources of survey error.[2] Every survey contains survey errors, most of which cannot be totally eliminated within the limits of our resources, and some cannot be eliminated even in principle in a *sample* survey.

The recognition that perfection is unrealistic brings us quickly to some practical questions:

- What are the potential sources of error in the survey we are planning?
- Which of these sources should most concern us?
- What reasonable steps can we take to reduce these main sources of error?

As in all other aspects of the design and conduct of the survey, decisions about how to handle sources of error must balance costs and other resources against the potential harm of not addressing an error source.

Some sources of error are more damaging than others. Before we can assess these sources competently for our particular study, we need to understand sources of survey errors in general. This understanding rests on two concepts: variance and bias. These two error sources, taken together, are used to assess total survey error. For our purposes, only a general understanding of these concepts is necessary, so we will approach them mainly by example.

Variance and Bias

Variance refers to the differences measured in repeated trials of a procedure. This is a useful concept, even though, in most surveys, we do not actually perform repeated trials. The most common example of variance, already introduced, is that of sampling variance. Recall that if we select a sample of size n and take a measurement on it (i.e., ask each respondent a question such as "How old were you on your last birthday?"), we produce one sample estimate of a population parameter, in this case, average age.[3] Then, if all other aspects of the survey are unchanged, we select a second independent sample of size n, take the same measurement, and produce a second estimate of average age. If this process is continued, we would expect to see variation in our estimate of average age from one sample to another. In other replications of a survey, again holding all procedures constant, we might expect to see random variations in such things as the percentage of respondents who can be contacted for an interview or in the number of refusals to answer a question about personal income. In any particular trial, the magnitude of the variation may be higher or lower than the average across trials.

A similar effect can result from the survey question itself. Imagine asking a sample of respondents, "How many times per month do you go shopping for groceries?" This will produce some sample estimate, say a mean of 3.4 times per month. Now, assume that a week later that same set of respondents is asked the same question again. Many respondents will report the same number as when first asked. But some respondents will report a different number, a bit higher or a bit lower. This may be a result of various factors, including simply thinking about their answers more or less carefully the second time; but the point is that just as we might produce varying estimates from different samples, so we might produce varying estimates from different administrations of a survey question to the same sample. In both instances, if the differences are random we consider them as a source of variance.

By contrast, bias occurs when a measurement tends to be consistently higher or lower than the true population value. In the example just cited, there might be a tendency for some respondents to report that they are older than they actually are, a consequence, for example, of ambiguous wording of the question. Assume we ask simply, "How old are you?" Most respondents will give their current age—that is, their age on their last birthday. But some others may decide, if they are approaching a birthday, that they should report the age they are about to become. It is unlikely that any respondents will report their previous age, even if they just had a birthday. So whatever misreporting occurs is not random, that is, it is not as likely to be in one direction as the other. Misreports are likely to produce higher than actual ages. The resulting sample survey estimate of average age will be higher than the true average. In this case, we would say that this measurement is upwardly biased.[4] Note that this source of error exists in addition to the variance.

More often than not, especially in small-scale research, we do not (or cannot, within our resources) produce empirical measures of the various sources of error in our particular survey, with the major exception of sampling error. Our design decisions concerning the nonsampling error sources are driven by the findings of other surveys and experiments where such measurements have been made. We use the more general (and hopefully robust) findings from such prior methodological research to guide us in rooting out the *likely* sources of error in our study.

In a discussion of the state of survey research as a science, Groves (1987) identified the two approaches to the issue of survey error as *measurement* and *reduction*. That is, there are those "who try to build empirical estimates of survey error and [those] who try to eliminate survey error." Logically, it would seem that the researcher should be equally concerned with both measurement and reduction and that the researcher's efforts would be directed, based on empirical estimates, to reducing the main sources of error. This is generally not the case (for reasons not discussed here). Nevertheless, the discussion to follow focuses mainly on the reduction of error during data collection, providing guidelines for identifying sources of error and suggesting steps to reduce their effect on the study's results.

Measures of Survey Quality

Survey error arising during data collection can potentially be serious. For example, if nonrespondents differ from respondents, the survey estimates will be biased to some degree. Of course, typically we don't know if respondents differ from nonrespondents on the survey measures because, by definition, we have no data from nonrespondents. In the absence of survey data we often look at indirect indicators. For example, in a general population survey we know from census data the expected distribution of some demographic characteristics: age, sex, race, education, and so forth. If, as is normally the case, we collect some of this information in the survey, we can compare our survey *respondents'* demographic characteristics to the census. Suppose we find that (a) our respondents underrepresent some demographic groups and (b) some of the substantive survey questions tend to be answered somewhat differently by members of the underrepresented groups compared to nonmembers? Would we be concerned that our estimates may be biased against the underrepresented groups? It is important to note that if a group is underrepresented, then some other group must be overrepresented. Think about why this is true.

The relationship between demographic characteristics and substantive variables, if any, may not be known until after data collection is done and analysis is under way. Our only insurance against these potential biases is a good response rate. The response rate is the percentage of eligible sample units for whom interviews are obtained. This is called the *unit* response rate. We will have more to say about this later, including how it is computed. A similar measure is the *cooperation rate*, which is the percentage of sample members who are interviewed, divided by interviews plus refusals. Consider how these two rates differ.

Ideally, of course, we want to obtain answers from all selected respondents to every questionnaire item. We know that deviations from this objective occur at two levels, the unit, by which we mean a person or household (although it can also be an institution such as a business or school, if that's the survey population), and the item, which is an individual question in our questionnaire. Data that are missing at either the unit or item level can pose potential problems for the quality of our survey estimates. If we fail to obtain any information from some respondents, and for others fail to obtain complete information, our estimates and other analyses may be distorted, sometimes quite seriously.

Unit response rate is the main, and most widely accepted, indicator of survey quality. Of course, respondents who agree to the interview may not answer all the questions. They may refuse to answer particular questions, or inadvertently skip some items, causing *item* nonresponse. While the concerns about item nonresponse are the same as for unit nonresponse, this source of error is usually a less-serious concern. Typically, respondents who agree to the interview answer all, or nearly all, the questions. However, if a survey asks sensitive questions (e.g., sexual behavior or illegal acts), or questions that many respondents simply find too difficult to answer, item nonresponse can become serious. Item nonresponse is usually concentrated in just a few questions.

Both types of response can be affected by interviewer performance. Interviewer behaviors affect respondents' willingness to participate in the survey, and can affect their willingness to answer particular questions.

Interviewers can also impact the reliability of respondent answers. That is, the interviewers themselves can be a source of variability in the survey results, for example by misrecording answers to open-ended items or by being inconsistent in handling respondents' questions or problems. Hence interviewer training and supervision are crucial to effective data collection.

Interviewer performance is seldom measured quantitatively—separately, that is, from the aggregate response rates. Yet their performance is an undeniable potential source of survey variance and bias.

Unit Nonresponse

We are concerned about unit nonresponse because it occurs for reasons that often result in patterns of missing information. For example, suppose that in the crime survey, at the unit level, sample members who are male, or who have less education, or who are elderly living in suburban areas tend to be less likely to cooperate. If such sample members, on average, have different attitudes or experiences than survey cooperators, then our results, which purport to represent the state's *entire* adult population, will be affected. For example, men may be less willing than women to consider sentencing alternatives; people with less education may rate the job police are doing lower than other respondents; or the elderly may be more likely to avoid going certain places because of concerns about crime. Each of these possibilities is speculation, but such patterns are often found in survey results.[5] To the extent that opinions and behaviors differ by subgroups, their overrepresentation or underrepresentation will affect results.[6]

In the 1970s, as telephone surveys became the predominant means of general population data collection (outside the federal government) Dillman (1978) pointed out the need to examine each step of the survey process in detail, as a contributor to the final success of the survey, particularly response rates. It is useful to return to that advice, taking into account both the new tools and obstacles in conducting surveys today.

If we consider, in sequential order, all the components of survey implementation that may affect the participation decision, we realize that some important factors can easily be overlooked or not given sufficient attention in our planning (Exhibit 9.1). For example, we note that some components that occur before an interviewer ever reaches the respondent may provide information that affect the decision whether or not to participate.

Sometimes a telephone survey is preceded by an advance letter to those sample households with published numbers. As part of random respondent selection, the interviewer will often talk to another household member before reaching the respondent. That person's impressions can affect access to the respondent, the respondent's willingness to participate, or even the respondent's initial understanding of our study. Even if we don't speak to anyone in the household, we may leave a message about the survey on a home recorder. In some studies we may include in the advance letter or leave a message providing a 1-800 number or a Web site URL where information about the survey is available. People who chose not to answer the phone may nonetheless have noticed our phone number or organization name on a caller ID system, perhaps many times.

Each of these steps that precede the interview may affect cooperation and/or data quality. What do we want to do about them? It helps to think this through by putting yourself in the potential respondent's position. This will be a useful exercise. Before proceeding, write a detailed outline of the data collection steps in a telephone, mail, or Web survey. (Exhibit 9.1 below will help you do this.) Then discuss with a colleague what you think could be done at each of these steps to affect a respondent's willingness to participate. One way to do this exercise is for your colleague to "play" the respondent and you "play" the researcher. At each stage of the survey, describe how you plan to carry it out; for example, what information will be in the advance letter, whether or not you will leave a message on answering machines and what it will say, how you will describe the survey to the first person you speak to in the household, and so forth. After describing each step of implementation, your colleague-respondent tells you how he would react to it. Would the planned approach have a positive or negative effect? What questions or doubts might it raise in your colleague-respondent's mind? Each time you get a negative reaction, consider an alternative approach and try that out on the "respondent." You will find that, without any special expertise, if you and your colleague try thinking through the process in simply a commonsense manner—*but from the respondents' perspective*—it will produce many ideas, concerns, and insights about how best to conduct data collection.

Some options available to *prevent* unit nonresponse apply to both interviewer and self-administered surveys. These options include designing interesting, logically organized, and nonburdensome questionnaires; using effective devices to provide information about the survey, such as advance letters (or e-mails) and well-crafted introductions or cover letters. Other crucial design factors apply only to one mode or the other.

Exhibit 9.1   Factors Affecting Unit Response

- Prior notification about the survey
- Efforts to reach the respondent
- Initial contact and respondent selection intermediary (gatekeeper) direct
- Requesting participation
- Follow-up efforts
- Refusal conversion

For in-person and telephone surveys, it is of critical importance to carefully train the interviewers. In self-administered surveys, respondent instructions must be absolutely clear and easy to follow.

Recent Increases in Nonresponse

Since the mid-1990s several societal and technological factors have affected survey data collection, especially for general population surveys. In some instances, new technologies, such as the World Wide Web and other computer-assisted data collection methods, have created potential opportunities for low-cost data collection; in other instances, technologies such as call blocking have introduced serious difficulties into survey data collection. An increase in the proportion of telephone numbers used solely for Internet access, fax machines and, to a lesser extent so far, cell phones have made sampling household voice telephone numbers more difficult and expensive.

The rise in the volume of telemarketing which many potential respondents find difficult to distinguish from legitimate surveys, is also a serious problem. The practices undertaken by many households to avoid continual bombardment by sales calls have had a detrimental effect on legitimate surveys.

The decline in telephone survey response rates is the most measurable manifestation of these problems. Compared to a decade ago, or even to the last 4 or 5 years, respondents have become more difficult to reach and less willing to participate when contacted. In a review of nonresponse trends in several federally sponsored survey, Groves et al. (2004) show consistent, although not dramatic increases in nonresponse. It is generally agreed that nonfederally sponsored surveys do less well.

So far we do not know how much falling response rates have reduced the quality of surveys and the confidence users should have in their findings. But there is little doubt that continued deterioration in response rates could eventually be very detrimental.

Survey researchers have responded in many ways to halt this trend. While we will not separate new strategies to maintain response rates from procedures that traditionally have been used, it is important to understand that the allocation of survey resources to those aspects of the survey most closely related to obtaining response has generally grown.

Item Nonresponse

Data may be missing randomly; that is, any question is as likely to be missing as any other. Randomly missing data may be a result of, for example, interviewer mistakes, respondent error (e.g., mistakes in following mail questionnaire skip patterns or instructions), or even because of coding errors. If the amount of randomly missing data is not too large, our results should not be greatly affected. Certainly, if missing data do not exceed a few percent, we are not too concerned about its effects. In such a case, we should not expend many resources to reduce or eliminate the problem of randomly missing data. Unfortunately, when data are missing, most of the omissions are not random.

Even items that are missing as a result of errors by respondents or interviewers are likely to have a pattern. In self-administered surveys, respondents are more likely to skip or not answer questions that are ambiguous, sensitive, or difficult, or that are preceded by an unclear skip instruction. Interviewers may inadvertently encourage nonresponse (or affect answers) to questions that they themselves are uncomfortable asking. We need to be aware of such potential problems during questionnaire design and testing, as well as in interviewer training.

Our Approach: Decisions and Procedures

Next we will review ways in which specific sources of error arising from data collection are addressed. Our approach focuses on two types of decisions, those related to design and those related to procedures for implementing that design. By design, we mean the selection of a data collection method, development of the interviewer training protocol and the data-collection plan (including callback and refusal–conversion efforts). These design decisions, while not irreversible, are, once made, relatively set. In the course of the study, we cannot easily switch from mail to telephone, redesign our training plan, decide to offer cash incentives to participate in the survey, or add a special round of refusal conversion. These decisions, made at the outset, define the main features of data collection—almost always the most expensive stage of the survey—and changing them will usually have serious cost consequences.

We define *procedures* as components of the study's conduct which, although established at the outset of the survey, we can alter or adjust after the study is underway. *Procedures* require ongoing monitoring and microdecisions in the course of the study, such as dealing with problematic interviewer or respondent behaviors and handling problems with gaining cooperation.

To some extent, the distinction between design and procedures is artificial. Still, we think this division will provide a useful framework for separating the main design decisions that must be fixed fairly early from the ongoing microdecisions that occupy most of our attention while conducting survey data collection.

As is true of all our decisions about a study, data collection procedural choices are made within the constraint of our budget; each option has some effect on cost, which is interpreted here in a broad sense, encompassing both money and other resources, such as available time of the researcher or unpaid (volunteer) assistants or classmates. Even if we have available to us certain "free" materials and services, they are usually limited, thus requiring decisions about how best to use them. For example, although a university faculty member conducting a survey may not have to pay for a secretary to type the training manual, the secretary's time is not devoted solely to the project. So a choice may have to be made about whether the available time is best used for typing the manual, handling advance letter mailing, or performing other tasks. Finally, sometimes it is necessary to reallocate resources during the survey to address particular problems.

Whether interviews are obtained in person, by telephone, by mail, via the Internet or other means, the data collection process requires routine tasks, such as mailing, dialing sample phone numbers, setting up sample files, and tracking the sample results. These are fairly simple procedures, but they can introduce error. We need to be sure that phone numbers are dialed accurately and that sample tracking accounts correctly for all the sample released for data collection.

Routine processes can be set up to ensure that these largely clerical tasks are done carefully and do not introduce more than trivial error into the survey. These components of the survey are largely record-keeping tasks. Many computerized data-collection tools, such as CATI (computer-assisted telephone interviewing), have utilities that will help with much of this. We will want to think through the steps that must be carried out and for each step, develop a record-keeping form. This mundane procedure will ensure that we do not find ourselves in such situations as omitting followup mailings to some nonrespondents (or wasting resources mailing to sample members we have already interviewed or from whom we have already received questionnaires), not giving each phone number equal call attempts, or neglecting to rework soft refusals,[7] all of which can contribute to error and wasted resources. However, with moderate attention to such details, these sources of error can be reduced to triviality. The main claim on our resources during data collection will be activities that have the potential to reduce the more serious survey errors we described above: unit nonresponse, item nonresponse, and interviewer and respondent effects.

Interviewer-Administered Surveys

The respondents' decision to participate in a survey can be affected by a host of factors (see Groves et al., 2004 for a description of some ways to model these factors). We have little or no control over some of the factors, such as the survey topic, the survey sponsor, a respondent's predisposition to participate in *any* survey, or the social context in which the survey occurs. We try to make the topic appear salient and interesting and we emphasize the survey's importance; but there is only so much we can do without being misleading.

The factors affecting survey response that we can control are the survey administration procedures and, to some extent, interviewer behaviors. We try to design procedures so that each step (see Exhibit 9.6) is implemented to best effect. Yet, procedures such as a good advance letter, a well-written survey introduction, and even monetary incentives can all be undermined by interviewers who are not skillful in their interactions with respondents. A well-conducted survey must be concerned with both data-collection procedures and interviewer effects.

We first turn to data-collection procedures and then consider a number of interviewer performance issues, mainly in telephone studies. Although higher response rates are usually attained in in-person surveys, for cost reasons they are much less common than telephone studies. We include some discussion of in-person surveys because there are exceptions, such as for special populations. In-person data collection also sometimes finds a place in multimode surveys.

It is important that our resources be focused where they will be most effective. In either type of interviewer-administered survey, efforts to address unit nonresponse are often labor intensive and therefore expensive. This is true for both telephone and in-person surveys, though the latter are considerably more costly. In either case, decisions about procedures to enhance cooperation will have an important impact on our limited resources. Our discussion mainly concerns telephone surveys, issues for any interviewer-administered survey, with additional remarks as appropriate, on points unique to in-person studies.

Administrative procedures and quality control in surveys are greatly aided by computer-assisted data-collection and data-management tools. The acronym CASIC (computer-assisted survey information collection) refers to the wide and expanding array of computer-driven technologies used to collect survey data.[8] The two major CASIC tools are CATI and computer-assisted personal interviewing (CAPI). Both methods, which have been in use since the late 1970s, allow interviewers to enter survey responses into a computer file while conducting a interview. The main advantage of these systems is quality control. CATI and CAPI systems handle questionnaire skip patterns automatically, reducing interviewer error; limit the range of values that can be entered for some items, for example, requiring two-digit entries for age in a survey of adults; and check some answers against others for internal consistency, for example, alerting the interviewer if someone born in 1970 reports graduating from high school in 1980. The value of these technologies is that they perform these functions during the interview, rather than at a later, data-cleaning stage, preventing costly recontacts of respondents.

Many of these systems also automatically handle some sample administration and interview scheduling tasks. Most professional and university survey organizations have some type of CATI system; CAPI is, like in-person surveys themselves, far less widespread.

If there is a CASIC system available for our study, then the time and cost to learn how to use the system, including programming the questionnaire into it, must be added to our budget and project plan. For a one-time study, except with the simplest systems, it is probably best to subcontract the programming rather than to try to learn it.

These tools can be very useful to track progress, examine data, and adjust some procedures as data collection progresses. For example, we can more readily make changes in household screening procedures in a computer-assisted environment if we think such changes will make screening more accurate or improve cooperation. Similarly, we can revise the callback plan if we think it can be made more efficient.

We can check the data for patterns of nonresponse. For example, are we having particular difficulties in certain geographic areas (such as central cities of metropolitan areas) or with particular subgroups (such as young males)? By comparing our data to census distributions (in total for in-person studies or just for telephone households), we can get a very good idea of such disproportionate response. If we find such patterns for missing units, we may want to shift our efforts to those underrepresented groups, by either allocating more of the total interviewer hours to these groups or assigning interviewers with particular skills to make the contacts. If, for example, it appears to be more difficult to complete interviews in certain locations, we might increase the proportion of those case assignments given to the more experienced interviewers. If there is a higher refusal rate among men, we might assign those cases to interviewers that are among the better refusal-conversion specialists. Another pattern that can emerge is that particular interviewers are contributing disproportionately to unit nonresponse. If that is the case, retraining or reassignment may be necessary.[9]

It is important to keep in mind the power and flexibility that these CASIC technologies as we discuss the administrative steps in data collection.

Advance Notification

Prior to the start of calling (or of in-person visits) we may want to send an advance letter informing sample households that they have been selected for our survey, about the survey topic and sponsor, and the reason the survey is being conducted. Such a letter should assure confidentiality and provide a phone number (or possibly the URL of a Web site) that respondents can contact for more information about the study. The letter should be brief and, in addition to providing basic information about the survey, explain to the potential respondent why the survey is important.

Returning to the Maryland crime survey, how might we construct an advance letter along these lines? The word "construct" is used intentionally. To form the letter, we want to assemble a set of components, each of which addresses a specific factor that may affect the decision to participate. However, if respondents do not read the letter, it will not serve its purpose. To that end, we strive to keep the advance letter brief and easy to read. We must include only those points that will most likely affect response, and to express them concisely.

This letter is composed of several brief sections (Exhibit 9.2). Some of the sections simply describe the project, others stress the project's importance, some explain who is sponsoring it and what will be done with the findings, and others state where to get additional information. As an exercise, consider each sentence in the letter one at a time: What is the purpose of the sentence? How important to gaining cooperation do you think the sentence is? Again, try to read it from the perspective of a respondent. If you received this letter, how would you react? Would it help persuade you to participate in the study?

Advance notification is clearly a cost we can choose whether or not to incur. Whether it is worth the cost and time is difficult to know for sure. Also, letters can be sent only to those households with listed telephone numbers. (The vendor from whom you purchase the sample can also provide addresses. If you select your own sample, it can be sent to a vendor who will run in through a commercial database and return listed addresses to you.)

We recommend sending advance letters whenever possible. The letter adds to the legitimacy of the survey and helps to differentiate it from marketing, which is no small issue. It can also serve an additional purpose. For those households with unlisted telephone numbers that raise questions about the survey's legitimacy, the interviewer can offer to send the letter if the potential respondent will provide an address. Having already prepared such a letter will speed up this process.

Reaching the Sampled Respondents

The effort and care we have taken to design and draw a probability sample means little if many of the selected respondents are not contacted

Exhibit 9.2   Advance Letter

Jones family
2189 Cedar Rd
Adelphi MD 20799

Dear Jones family,

In a few days your household will be called and ask to participate in an important study. The University of Maryland is conducting a survey for the state of Maryland's Summit on Violent Crime.

Your household has been randomly selected for this study. The survey is intended to be representative of all types of households in Maryland, including those that have been touched by crime and those that have not.

All your answers will be kept strictly confidential and the survey results reported only in group form.

The Summit on Violent Crime will use the survey results to help plan crime prevention needs around the state.

Your participation is voluntary, but very important to the success of the study. The interview will take about 10 to 15 minutes. If the interviewer happens to call at an inconvenient time, she will be happy to call back at the time that works best for you.

In you have any questions about this study, please call us toll free at 1-800-314-9090 or visit our web site at www.stopcrime.umd.edu

Thank you in advance for taking the time to help in this important project.

Sincerely,
Robert Ellis, PhD
Department of Research Design

---

and interviewed. We have noted the potential problems resulting from missing data and touched on the contributions of interviewer and respondent behaviors to this problem. Now we turn to the use of callback procedures. Numerous studies show that repeated callbacks have a major effect on increasing responses from the selected sample regardless of the data-collection method. Callbacks are the single most-effective strategy in reducing nonresponse.

Whether the sampling frame is random-digit dialing (RDD), a list, or some combination, after the first round of calls, the sample is sorted into several groups. These groups include some interviews and identified nonhouseholds that need no further attention. There will also be some early refusals, telephone numbers whose status (household or not) we have not determined (ring-no-answer numbers), and some households in which the respondent (or house-hold informant) could not be interviewed (at least on the first attempt) because of problems such as difficulty hearing, illness, or languages other than English. In addition, we will have a large number of noncontacts, which include reaching answering machines and finding out that randomly selected respondents are not at home or are busy.

For each of these initial dispositions of the sample numbers, we must decide on a followup callback strategy. How well we do this has a major effect on our success in reaching and interviewing selected respondents and on the costs necessary to do so.

In telephone and in-person surveys, we have some information about the sample units that did not result in an interview on the first call. Even if this information amounts to no more than when the call attempt was made, we can use it to fashion our callback strategy. But we have this information only if interviewers record it carefully.

Interviewers must be trained in the use of forms and procedures for recording, and aggregating, in sample status reports, the result of each call attempt, that is, of each time a sample number is dialed. Regardless of the outcome, the interviewer codes what happened on the call. This kind of coding can be quite elaborate, but we recommend making it no more complex than is absolutely required by the study. The essential call results include completed interview, initial refusal/partial interview, final refusal, noncontact, and other nonrespondent problems. For each of these dispositions the date, time of day, and day of week should also be recorded.

The results of initial calls to each number in the sample determines how we will subsequently "work the sample," meaning whether and when we will schedule additional call attempts and, to some extent, which interviewers will make the calls. How well we do this has a major effect on both costs and nonsampling errors. Because each call to a sample number has a cost, the objective is to reach a disposition for each number with the minimum effort and to identify eligible sample members and complete interviews with as many of them as possible. Costs are affected because a large part of the total survey budget is for data collection, and calling the sample numbers represents—after the actual conduct of the interviews—the next major portion of data collection costs.

Exhibit 9.3 shows the distribution of initial call results (after one call to each number) and the actual final distribution for the University of Maryland Survey Research Center 1992 State Crime Survey. The first column is the equivalent of our first sample status report. A large number of nonhouseholds are identified after the one call, but more do turn up in the following calls. Initially, we are not sure whether a large group of telephone

Exhibit 9.3   Maryland Crime Survey: Disposition of the Sample after 1 Call Attempt and after 20 Attempts

	After 1 Call		After 20 Calls	
Total sample	1,816		1,816	
Nonhouseholds	545		702	
Household status unknown	436		53	
Households	835	100%	1,061	100%
Interviews	181	22%	824	78%
Refusals	54	6%	164	15%
Noncontacts	527	63%	49	5%
Problem households (language, problems hearing, illness, etc.)	73	9%	24	2%

numbers are households or not. This category is greatly reduced (although not to zero) over the subsequent calls. In more than 60% (527 of 835) of the identified households, the selected respondent could not be interviewed on the first call attempt. However, by the end of the study, the noncontact rate is reduced to 5%.

Taking each disposition in turn, we consider what types of information we are likely to gather on the first (and subsequent) call attempts and how to use it. Some sample numbers almost immediately drop out because they are not part of the eligible household population: businesses, nonworking numbers, disconnected numbers, government offices, and so forth. However, even some of these numbers may require more than one call.

The second group that is quickly finalized are respondents who are easily contacted and agree immediately to the interview.[10] On average, these are respondents who are home frequently and who quickly comply with the request for an interview. Those at home more often are, as a group, more likely to be elderly, retired, unemployed, and, to a lesser extent, women. While all of these are eligible sample members, clearly they are not a random sample of the population. Using these samples would risk producing very biased population estimates. There are several tempting, but incorrect, procedures that would result in such poor samples. For example, we could draw a very large sample of phone numbers, begin calling and stop when we reached the target number of interviews, leaving the rest of the sample unworked. Or we could make one or two call attempts to each number in a selected sample. Both these approaches would skim off those respondents most available for interview, and both would result in overrepresentation of the demographic groups listed above.

Another factor that leads some respondents to immediately agree to do the interview is the survey topic. Respondents who are interested in the survey subject are less likely to put off the interviewer. In the Maryland crime survey, these respondents may be those who have been crime victims themselves or, for other reasons (e.g., because of their profession or the experiences of friends) have strong feelings about crime issues. Again, although they are eligible sample members, a survey that included only these types of respondents would probably provide misleading results. Both the easily reached and the quickly compliant require no further decisions or effort; both also illustrate the nonrandom nature of convenient subgroups of our sample. Now that it is clear that the transition from sampled household to completed interview is not a random process, we must decide how to direct our efforts (resources) to maximize the proportion of the sample that is interviewed and to weed out the ineligibles most efficiently.

The sample cases remaining after the first round of calls fall into two categories: households and household status unknown. As we work the sample, these same patterns recur: some respondents are relatively easy to reach and interview; others require greater efforts either to contact or to persuade, or both.

As discussed in Chapter 7 on sampling, we typically begin data collection by releasing a random subsample of numbers for calling. After estimating the interview completion rate (interviews divided by finalized sample) based on this subsample, we release additional random subsamples to attain the total number of interviews we want for the study. When these telephone numbers are released to the interviewers, the call-results pattern is usually similar though not identical.

Number of Contacts. The main factor in successfully contacting a high proportion of selected sample members is simply the number of contact attempts. Surveys that rely on a single contact attempt are likely, in almost all cases, to have serious nonresponse bias. For telephone surveys, the number of attempts typically ranges from 3 to 20, and for mail studies from 2 to 4 (Groves, 1989). We recommend no fewer than five attempts for general population telephone surveys and at least two mailings (each including a questionnaire and a self-addressed, self-stamped return envelope) for mail studies. If, during data collection, we realize that the planned level of effort is not producing the anticipated response rate, we may need to make some adjustments in our callback plan. Unfortunately, because of cost, we do not have the latitude to make many additional contacts. But adding one or two additional calls for all noncontacted sample members is often useful. This may be especially true if, by examining past call results on each number, we can focus those calls on days and times not yet covered.

Answering Machines. In recent years the ownership of home answering machines has greatly increased. Early research (Oldendick & Link, 1994; Triplett, 1994) showed that most of these households can be reached with proper scheduling. Weekend mornings are particularly effective times to contact these households, and they do not appear, once reached, to be any less cooperative than households without answering machines. In the crime survey, after one round of calls, approximately 13% of the identified households were dispositioned as answering machines. By the end of the survey, this number was reduced to approximately 2%. Although this calling strategy is still recommended, there is some evidence that it is becoming less effective. In a large, ongoing immunization survey, conducted for the Centers for Disease Control, independent samples are selected for every calendar quarter. In 2002, the percentage of answering machines at the end of data collection began to rise.

Appointments. It is not uncommon for respondents to telephone surveys to request that they be recontacted at a more convenient time and are willing to set an appointment for the interview.[11] It is extremely important that the interviewing effort be organized to ensure that such appointments are kept. Missing them can easily turn a willing respondent into a noncooperator.

Call Scheduling. To properly and cost-effectively work the sample, interviewers must keep an accurate and complete record of the result of each call attempt. This is done through the use of a simple form that the interviewer completes immediately after calling each number. This record allows the interviewing staff manager to look for patterns as to when and when not to attempt additional call attempts for each household. If, for example, a particular household has never answered on weekday late afternoons, it makes sense to shift the next few attempts to later in the evening or to the weekend. Household informants, once reached, can often provide information about the best time to reach the randomly selected respondent. This information is recorded on the same form.

Using a CATI system that has a call scheduling feature is more efficient than manual methods. However, it will require that some decisions be made and entered into the CATI system. You will need to decide when the first call should be made—day or evening, weekday or weekend—and, if that call is a noninterview, how subsequent callbacks should be scheduled, depending on whether the call result is a ring-no-answer, a refusal, an answering machine, or something else.

In general, for an RDD survey, it is efficient to make the first call during the day because doing so facilitates eliminating businesses and other nonhousehold numbers. After that, weekday evenings and weekends are when people are most likely to be at home. It is important that callbacks be spread over different days.

Identifying Bad Numbers. We will have difficulty determining the residential status for some telephone numbers even after several call attempts.[12] Because of technical features of the telephone system, a phone number that simply continues to ring each time it is called may or may not be a residence and may or may not be in service. Repeatedly calling a nonresidential number may not only be wasting effort and money but, in the case of the Mitofsky-Waksberg sample design (two-stage RDD), would be costing us the opportunity to replace the number. There are three things we can try: first, schedule a few calls at earlier or later times than usual to try to catch people with, for example, odd work schedules. The few calls at odd times can be useful in bias reduction if the numbers turn out to be residences and an appointment for interview can be set up. Such respondents may be different, both demographically and substantively, than those with more regular schedules. For example, people who work nights may, on the whole, have different responses to some of our questions on crime than do other respondents. Second, when possible, look up the number in a reverse directory.[13] Third, try the local phone company for information. Unfortunately, the usefulness of this last option varies greatly by the particular company's willingness to help. But it is an inexpensive option to try.

Cell phones and computer and fax lines are rapidly increasing. Although most cell phones are usually assigned to different exchanges than land lines, this is not invariably the case. Also, some respondents may have calls forwarded from a land line to their cell. At this point in the state of technology and sampling frames, all we can do is to train interviewers to be alert for this possibility. Computer and fax lines are more easily recognized and coded, for our purposes, as nonresidential numbers. If a respondent reports that a line is used for voice and computer or fax, classification as residential or not depends on what the respondent says is the *primary* use of the line.

Reworking Refusals. A very important procedure for telephone surveys is refusal conversion. Respondents who refuse to be interviewed initially can often be "converted" on subsequent attempts. While there is little research on this phenomenon, we suspect that conversion is possible partly because the reason for many initial refusals has nothing to do with the survey itself. Keep in mind that while our survey is very important to us, it might mean little to the typical respondent. Some people, for example, caught at a bad time, will take a moment to set an appointment or just tell the interviewer to try some other time, while other respondents in the same circumstance will refuse, but when recontacted, some will agree to the interview. Professional survey organizations can, using specially trained staff, typically convert 30% to 40% of initial refusers. While the nonprofessional will probably not achieve such rates, a nontrivial number of first refusals can likely still be turned into interviews. In addition to reducing the refusal rate, this procedure can also reduce bias if the initial refusers are generally different in some respects from the initial cooperators.[14]

Typically, we will want to let some time pass between the initial refusal and the attempt to convert. If the refusal was a result of a "crisis" in the respondent's household, perhaps in several days it will be resolved. Also, many respondents will not even recall the original contact, so we may not want to even mention it, but, rather, start fresh. The approach we take for trying to convert a particular refusal depends, in part, on what happened the first time. For this reason, it is extremely important for the first interviewer to note the reason, as far as can be determined, for the refusal—the more detail, the better. The record should note the sex of the respondent and indicate, for example, that the person seemed very elderly. Keep in mind that the followup call might reach a different person in the household, if we had not gotten as far as random respondent selection on the first call.[15] Still, it is the circumstances of the first refusal that should be noted. If, for example, a person refused because she was about to leave for work, it would be foolish to make the conversion attempt at the same time of day as the first call, even though calling at that time increases the chances of reaching the respondent. Similarly, if a respondent refused initially because of a perceived sales purpose, the refusal converter should certainly be prepared to counter this perception quickly if it is raised in the followup call.

Problem Households. In some households, after reaching the randomly selected respondent, we find that the interview cannot be conducted. Some respondents do not speak English at all or not well enough to be interviewed. If we are doing the survey in an area with a large non–English-speaking population (e.g., Miami), we would risk serious coverage bias by not making provisions for interviewing in another language, in this case, Spanish. However, in most small-scale research, we will not have this capability. Nevertheless, these households are still part of the defined population and must be counted as eligible households in our response rate calculations. Similarly, people who are too ill to be interviewed, or who have some physical or mental disability that prevent them from either hearing, understanding, or answering the questions, are also lost to us.

Minimizing Item Nonresponse

Like unit nonresponse, unanswered questions (item nonresponse) typically do not occur at random.[16] Respondents may be reluctant to answer particular questions (e.g., sensitive items such as whether or not they carry a weapon for protection) or may have difficulty answering others (such as how likely they think it is that in the coming year someone in their household will be a crime victim).

The second class of items that have a higher likelihood of nonresponse are questions that are difficult to answer or require checking records. For example, in a health survey it might be of interest to know how much the respondent spent on prescription drugs in the past year. For many respondents, this will be easy, because the answer is zero. For other respondents, who had many such purchases, this could be very difficult. When faced with such a question, some of those latter respondents will try to answer or hazard a guess; it is quite likely that others will simply say they can't recall. Some items, like "total household income from all sources," may be both somewhat sensitive and difficult for certain respondents (e.g., respondents in households with many sources of income).

Similarly, factual questions that request great detail about issues, behavior, or events that are of low salience to the respondent may require more effort than many are willing to expend. While most people will know how many doctor visits they had in the past 6 months, many people will not know how many times they went to a grocery store.

The solution to many of these problems is in questionnaire design. Ask sensitive questions later in the survey, after some rapport has been established between the interviewer and respondent. Preface such questions with reassurances about confidentiality and/or about the need to ask such questions. Provide categories for difficult numerical questions rather than asking for exact figures.

During pretesting note whether particular items seem to have an unusually high proportion of Don't Know or Refused responses. Some item nonresponse can, of course, result from simple errors in skip patterns, another issue to check carefully prior to and during pretesting.

Other reasons for item nonresponse include poorly written questions. If respondents cannot make sense of the question or have to work to do it, many will not answer it at all. Interviewer behaviors, discussed below, can also affect the item response rate.

Interviewer Effects

After we have done all we can by way of instrument design, advance notification, and setting up data collection procedures, gaining cooperation is in the hands of the interviewers. Moreover, interviewers can affect both the willingness of respondents to answer particular questions and the quality of those responses.

Imagine that, for the crime survey, an interviewer contacts a household and selects, at random, an adult who says, "I really don't have time for this. I was just going out. Why don't you talk to my wife. She keeps up with news and politics more than I do anyway." The interviewer responds with, "OK. She'll probably really like this survey." When the wife comes to the phone, the interviewer starts to read the introduction and the woman interrupts with, "Why is the university doing a poll about crime?" "Well," the interviewer responds, "I'm not too sure, but it probably has something to do with the governor's reelection campaign. You know, to show he's serious about getting criminals off the street." Eventually, the interviewer gets to the question, "In the past year, would you say that the VIOLENT crime situation in the STATE has gotten better, gotten worse, or stayed about the same?" The respondent says, "My daughter tells me she never goes out at night anymore since her next-door neighbor was mugged right outside his house." The interviewer responds with, "So I guess you really think crime's gotten worse."

Each of these interchanges shows how an interviewer can undo the careful design and procedures leading up to and including the interview.[17] First, the random respondent selection procedure is abandoned for convenience; then, the introduction is cut short and a personal comment is interjected that may affect the respondent's reaction to later questions. Finally, instead of probing for a response that matches the answer categories, the interviewer infers the answer from the respondent's comment and puts words in the respondent's mouth. This not-very-farfetched example shows how easily, even naturally, interviewers can affect the quality of the survey.

In-person surveys are subject to more potential interviewer effects than telephone studies. The interviewer's physical characteristics can influence respondent behaviors. For example, if the subject matter of the survey includes racial issues, the race of the interviewer might have an effect. Interviewers' facial expressions and eye contact, which are not pertinent on the phone, may become issues for in-person surveys, particularly those dealing with attitudes or sensitive behaviors. A major difference between the two data-collection modes, as far as interviewers, is the amount of monitoring and supervision that is possible. In-person interviewers, whether conducting household or special population surveys are much more on their own than interviewers in a centralized telephone facility.

We will spend a good deal of our resources in efforts to control interviewer behaviors through training, monitoring, and, most important of all, showing them how inadvertent, well-intentioned, actions can be detrimental to the research effort. We must keep in mind that it is the interviewer we depend on to carry out crucial aspects of the survey design. Our job often begins with interviewer selection.

Interviewer Selection

Not only the training but also the selection of interviewers is important. Interviewing is a skill that requires good reading competence, personal interaction ability, and, often, considerable persuasive know-how. We ask interviewers to contact strangers, usually without advance notice, request their cooperation for the interview, explain the purpose of the study, answer their questions about the study or about particular survey items, administer the interview according to instructions, and accurately record the answers. All this must be done while maintaining a professional demeanor, sometimes in trying situations. Not everyone is suited to all the tasks that interviewing requires.

In some instances, the researcher has no role in interviewer selection, for example, when the survey is a class project or a survey organization is hired to collect data. However, even in the case of a class project, it may be that not all the students need to or can properly conduct interviews. As we noted, unlike other aspects of the survey process, interviewing requires both an understanding of what happens in the interview process as well as the skills to carry it out. The best way to find out who has these skills is live practice after some training. One approach is to have everyone participate in the pretest and then decide who should conduct interviews and who should do interview monitoring, data entry, or some other project task.

Whether we are actually recruiting or just making assignments from the interviewers available to us, there are a few guidelines to consider. First, it is useful to have the prospective interviewers go through a structured test. Four areas should be covered: reading and following instructions; gaining respondents' cooperation; reading questions properly; and recording answers accurately. How much skill we can expect in the last two areas depends on how much prior experience, if any, the prospective interviewer has had. But we will find that even novices will differ in how instinctively they react to respondents' reluctance or how naturally, after even a few tries, they can read survey questions. The simplest way to screen possible interviewers is to have a few study-specific instructions available, along with a draft of the questionnaire. We should explain to each interviewer the purpose of the study and what we are going to ask them to do. After allowing each interviewer a few minutes to read the relevant instructions and the questionnaire, then a supervisor (or another student) plays the role of respondent, at first going along easily with the request for an interview and answering the questions without incident. On subsequent rounds, the "respondent" varies his or her behavior from this ideal, creating progressively more difficult situations for the interviewer. An observer (or the class as a group) grades the performance. If this test can be set up so that the two parties are in separate rooms actually communicating by telephone, a more realistic assessment of the interviewer's skills, as well as of the interviewer's telephone voice, can be made.[18]

Interviewer Training

The training of interviewers should cover both general interviewing skills and the specific requirements of the study at hand (Exhibit 9.4). The amount of general training required depends on the interviewers' prior experience; if possible, it should be conducted by an experienced trainer. The following discussion covers the most basic aspects of training; it supplements a training manual from a professional survey organization and should be followed closely.[19] Survey centers at public universities often will make their general interviewer training manuals available to other (not-for-profit) researchers for free or for a small fee.

The training sessions (and you should plan on multiple sessions) should focus on skill-building practices. At every opportunity, an effective training agenda will emphasize participatory exercises over lectures. We want

Exhibit 9.4   Key Subjects to Cover in the General Training of Interviewers

1. Gaining cooperation
    - Identifying or selecting the correct respondent
    - Explaining the purpose of the survey
    - Persuading reluctant respondents to cooperate

2. Administration of the interview
    - Getting started
    - Making the transition to the interview
    - Reading questions verbatim
    - Asking nondirective probes
    - Asking all questions and recording answers correctly
    - Following skip instructions
    - Recording answers correctly: open ends

constantly to point out behaviors, such as those illustrated at the beginning of this section, that seem natural in conversation and are well-intentioned but are inappropriate in an interview. Rather than simply laying down a series of rules—although we must do that, too—we must show interviewers why such behaviors are detrimental to the project. This point and others are best conveyed through practice and example, not lecture. One aid to doing this is to require that the interviewers read background materials before each training session.

Conducting the Interview

A central component of all aspects of interviewer training is role playing, sometimes referred to as mock interviewing. As we have noted, successful survey interviewing requires not so much conceptual understanding (though, of course, a measure of that is essential) as it requires execution of certain skills in gaining respondents' cooperation and properly administering the interview. Such skills are acquired only with practice. The less interviewing experience an interviewer brings to the task, in general, the more practice is necessary. It is also very important to realize that shortcuts on training are false savings. If the interviewers do not practice their skills in a training session, they will, of necessity, practice them with real respondents during the study. The preference is obvious.

In mock interviews, just as in the interviewer-screening process, interviewers take turns playing the role of interviewer and respondent. Each practice round is structured to address a particular set of skills. Practice continues in each area until the interviewer is comfortable, quick, and smooth in handling each situation. Remember that when a real respondent is on the line, there is no time to consult notes, hesitate, or back up and start over. Either an impatient respondent will end the interview or improperly administered questions will result in poor measures.

The exercises should follow the chronological order of interview administration, with at least one exercise to illustrate skill building for each stage of the interview. The number of exercises used and the amount of time devoted to each should be determined by the nature of the questionnaire, the interviewers' backgrounds, and how quickly they pick up the required skills. Whenever possible, after group instruction, having interviewers practice some of the exercises on the phone will lend realism to the training. The interviewers should know that the final step of the training will be to do a "live" interview with a real respondent. These practice cases should be selected from the study population but not from the actual survey sample. Of course, these final practice cases should be closely monitored.

The fundamental instructions to interviewers can be stated simply: Read each question and transition statement *exactly* as written, including all the response categories, without any deviation; ask all the questions, even if you think the answer has already been given; follow the questionnaire's skip patterns; and record answers carefully and, in the case of open-ends, verbatim. These rules are as easy to state as they are to break. This is particularly true of the maxim to read each question verbatim. Anyone who has conducted interviews has run into situations in which it seemed that adding a comment or changing a few words would "clarify" a question or "help" a respondent. The questions must stand on their own, without embellishments of any sort. While we cannot claim that every deviation from the exact wording of the question results in unreliable data, to open the door to invention completely undermines the foundation of uniform data collection. We must strive to develop questions that do not tempt interviewers to such circumvention.

One way to train for this skill is to have each interviewer, in turn, read a question or questions from the survey and have the others critique the reading for verbatim delivery and natural pacing, with proper pauses at punctuation marks and clear pronunciation. The interviewer's tone should be conversational and not sound like a reading from a book. Interviewers often want to skip this practice, feeling that they are literate and can read well enough. It is quickly evident that even well-educated interviewers, on first practice with a new questionnaire, often misread questions, go too fast for many listeners, and make occasional mistakes in pronunciation. These errors seem trivial until there is an irritated or confused respondent on the line. Then these "trivial" errors often lead respondents to hang up or to misunderstand questions. Administering survey questions is not a reading skill but a performance skill.

After the survey introduction and respondent selection, the interview proper begins. We have tried to design the questionnaire so that the first questions ease respondents into the interview; and in most cases the interview will proceed smoothly and uneventfully. But, as we noted in the discussion of behavior coding, there can be some problematic interactions between interviewers and respondents. These have three sources: the questionnaire, the interviewer, and the respondent. We have tried, through the questionnaire design and the training of interviewers to read questions verbatim, to minimize the first two problem sources; we now turn to the third.

Unit Response

The two key training areas are gaining cooperation (unit response) and conducting the interview (including item response). First, we focus on gaining cooperation. As we have noted, in recent years respondent cooperation in general population surveys has become more difficult to obtain. While the general approach to training interviewers to gain cooperation is largely unchanged, the amount of time spent on gaining cooperation practice exercises is more important and takes a larger share of training time than in the past.

Many respondents will not immediately agree to the interview. Sometimes they want to know more about the survey than what is contained in the interview's introduction. The interviewer needs to be ready to provide this information quickly, concisely, and clearly. A common practice is to have a one page "info sheet" listing the study's sponsor, it's purpose, the length of a typical interview, and a thumbnail description of how the results will be used. If, in the course of the pretest or early interviews, other respondent questions or concerns frequently come up, these should be added to the info sheet and circulated to all the interviewers. It is also very useful to provide a phone number respondents can call to verify the legitimacy of the survey.[20] A few respondents will want to be reassured about the confidentiality of their answers. But for most reluctant respondents, none of these will be the issue; they will simply not be interested enough to give their time.

The main appeal of a survey is its topic—if the respondent gets to that information. Many refusals in interviewer-administered surveys occur before the topic is mentioned. Survey introductions should be written and interviewers trained to broach the topic—if you think it is an interesting or salient one—as soon as feasible. Advance letters can, of course, help with this.

Topics naturally vary in their appeal to respondents. The crime survey will, in general, be more interesting to a larger number of potential respondents than a study about the public libraries. But even with an interesting, newsworthy topic, many people will not immediately agree to the interview. The interviewer must be prepared for this reaction. In a telephone survey, the interviewer has very little time to gain cooperation. Remember that it is easy for a respondent to hang up.

Two tactics for handling respondents' reluctance are (a) keep the respondent engaged and talking, and (b) address the specific reason the respondent does not want to do the interview. It is very important that the interviewer listen closely to the respondent's reason for not wanting to do the interview, even though many times this will have nothing to do with the survey but simply with bad timing. If the respondent is about to leave for work, is in the middle of watching a ball game, or is dealing with a personal problem at home, the alert interviewer will quickly back off from requesting the interview and try to determine a better time to call back. This approach would seem to be dictated by simple common sense, but it is not unusual for overly aggressive interviewers to push on in such situations, turning reluctance into refusal.

Even though the interviewer has read an introduction giving the survey sponsor and topic, some respondents will still suspect it is a disguised sales call or some other solicitation. This is partly because some respondents do not really hear the introduction. Remember that they were not expecting the call; also, many marketing campaigns are disguised as surveys. The interviewer should be ready for this and quickly repeat the sponsor, especially if it is a university or recognizable government agency.

A large number of respondents will simply say that they are not interested or that they don't have time. Although these responses sound different, they often amount to the same thing, a stock response to end unwanted calls quickly. The main thing the interviewer must try to do is keep the person engaged to avoid the imminent hang up. If the topic is a popular one, the interviewer should try to repeat it, still acknowledging that he or she listened to what the respondent said. For example, "Most people find they like giving their opinions about the crime problem once they get started. It won't take very long."

Some respondents will say they don't really know much about the issue, particularly if the topic sounds technical, like biotechnology or some aspects of the environment or the economy. The interviewer needs to assure such respondents that the interview is not a test of knowledge.[21] A response such as, "This is just about your *general* opinions. It doesn't require any special knowledge," will often suffice.

Finally, in some situations, the reluctance is not from the actual respondent but from a household informant. When we contact a household, any adult can answer the questions used for random respondent selection. If the selected respondent is then someone other than the person on the phone, the interviewer asks to speak to the respondent. At this point, the situation may involve a gatekeeper, that is, a person who answers the phone and does not want to let the interviewer talk to the respondent. The reasons may include some already discussed as well as others. The interviewer strategies are essentially the same, with the added component of expressing the desire to describe the survey briefly to the respondent. If the informant is still reluctant, calling back at another time at least gives a chance to bypass the gatekeeper and reach the respondent directly. Exhibit 9.5 lists suggestions for handling reluctant-respondent situations.

Item Response

Interviewer behaviors can affect people's willingness to respond. For example, a crime victimization survey might include a question about sexual assault. If the interviewer is uncomfortable asking the question, it might

Exhibit 9.5   The Reluctant Respondent: Typical Situations and Some Remedies

Bad Timing

- Don't try to push for the interview.
- Show understanding of the bad timing.
- Use judgment in deciding whether to even ask for a callback time.

Suspicion of Disguised Sales or Solicitation

- Repeat the sponsor and study purpose.
- Assure the respondent that this is not a sales pitch or solicitation; offer an 800 number, if available, or request that the respondent call collect for verification.

No Time/ Topic Not of Interest

- If survey topic is a popular one, focus on it; if not, focus on how quickly the interview goes.

Respondent Doesn't Know about Topic

- Focus on the opinion aspect of the survey; downplay any knowledge questions.
- Let the respondent know we're interested in what all people think about the issue.

Gatekeepers

- Ask that the respondent have a chance to hear what the survey is about.
- Call back to try to avoid the gatekeeper.

affect how the interviewer asks it, perhaps by reading it very rapidly to get past it and/or lowering his or her voice. Such behaviors will make some respondents more uneasy answering the question than they would be if the question was asked with the same pace and tone of voice as other questions.

Poorly trained or supervised interviewers can be even more blatant, with an aside to the respondent such as "you don't have to answer if you don't want to." There may be instances, as we have noted, when it is important to remind respondents of the confidentiality of their answers or even—at the beginning of a particularly sensitive section of the interview—that if they don't want to answer a particular item, just let the interviewer know. However, we want to control when this happens, lest some interviewers encourage respondents to skip questions they might otherwise answer.

In most surveys, item nonresponse is low. After agreeing to be interviewed, the typical respondent will answer all the questions. But interviewers must be prepared to handle reluctance. The interviewer has two main strategies for dealing with reluctance. First, let the respondent know why the question is necessary. Some respondents will want to know, for example, why some of the demographic questions (e.g., age, race, or income) are needed. Often, a simple response will suffice: "We want to compare our survey to the U.S. census to be sure it's representative." Second, remind respondents that all their answers are completely confidential and will never be linked to their name, address, or telephone number. If, after trying these approaches, the respondent still refuses an item, it is best to skip it and go on. Trying too hard to persuade the respondent to answer a particular question may lead to a breakoff and a lost case. Sometimes, if the item refusal happens early in the interview, it is possible to return to that question later, when the respondent may be more at ease with the interviewer and convinced of the legitimacy of the survey.

A question can be answered but unusable, making it, in effect, a nonresponse. For example, if the answer to an open-ended question (or to an *other specify* option in a closed question) is poorly recorded, the answer may not make sense or otherwise be uncodeable. If the questionnaire has open-ends, interviewers have to have practice recording answers verbatim.

Finally, just as errors in questionnaire skip patterns can cause some respondents to skip questions they should be asked, interviewers can make skip pattern errors when working from paper questionnaires. Again, practice exercises during training will help prepare interviewers to work accurately in the real-time situation of an interview. An important benefit of CASIC systems is to eliminate this type of nonresponse error.

Response Errors

We know that response error can occur when respondents misunderstand questions, cannot recall information, or otherwise have difficulty answering, or even purposely answer falsely. We try to address these problems during instrument design and testing.

The interviewer can also affect some types of response error for better or worse. Interviewers can affect how respondents interpret questions and can sometime influence respondent answers. Most often the effects occur in how interviewers handle some problematic respondent behaviors.

There are three main behaviors of respondents that interviewers must be trained to handle: (a) the respondent does not answer using one of the response categories; (b) instead of answering, the respondent asks a question; and (c) the respondent responds with a comment about the topic. The interviewer must deal with all of these carefully to avoid affecting the resulting data or having the respondent break off the interview.

For each of the problem situations, the interviewer must get the respondent to provide an appropriate answer that can be coded into one of the given categories, yet maintain rapport. Remember that the respondent is volunteering time and trying to be helpful. The interviewer knows that she needs an answer that fits a closed response category and that extraneous remarks cost time and money, but the respondent may view the interaction as something much more like a conversation, with digressions and asides being quite appropriate. It is the interviewer's task to get the respondent on track and keep him there—but tactfully. In doing this, it is paramount that all interviewers handle similar respondent problems in the same way and in a manner that does not affect the respondent's answer choice. It is for these reasons that we instruct interviewers about what sorts of things to expect in the interview and impose strict guidelines on permissible behavior in each situation.

In the mock interviews, "respondents" should take turns deviating from being good respondents while the interviewer practices responding to each situation. After each round, the interviewer's performance should be critiqued by the group.

When interviewers do not receive acceptable responses, their main tool is the probe, a question designed to elicit an acceptable response. Interviewers must learn to recognize which answers are satisfactory and which are not and to use probes that are nondirective, that is, they do not suggest an answer to respondents.

Consider the first crime survey question:

```
In  general,  would  you  say  that  the  crime  problem  in  YOUR
NEIGHBORHOOD  is  very  serious,  somewhat  serious,  not  very
serious,  or  not  at  all  serious?

    1.  VERY SERIOUS

    2.  SOMEWHAT SERIOUS

    3.  NOT VERY SERIOUS

    4.  NOT AT ALL SERIOUS

    5.  DK
```

If, the respondent's answer is "serious," the interviewer should probe by repeating *all* the response categories. "Would that be . . . (read categories)?" The interviewer should not infer from earlier responses, even to very similar questions, which category to code. Nor should the interviewer's probe make such an inference, as in "So would that be 'very serious'?" If the categories are such that a partial answer puts the respondent in one subset of response categories, then the probe can refer to that subset. For example, if the choices are "very satisfied," "somewhat satisfied," "somewhat dissatisfied," and "very dissatisfied," and the respondent simply says "satisfied," a proper probe would be "Would that be 'very satisfied' or 'somewhat satisfied'?" A bad probe would be one that did not repeat all the relevant categories. In each case, the respondent must always select the category. This practice should be followed even if the same problem occurs on several questions. Although most respondents will pick up quickly what is wanted, others have to be "trained" by the interviewer to be "good respondents," and the training must begin at the outset of the interview. Once the respondent is off track, returning is doubly hard.

In training sessions, interviewers should practice suggesting probes they would use in particular situations. It is also useful in practicing probes to note some tempting, but inappropriate, probes. For example, if a respondent answers "serious" to the question above, the interviewer should *not* say, "So, can I put you down as very serious?" Consider an open-ended question such as, "What do you think is the most important problem facing Maryland?" Suppose the respondent answers, "Drugs." A poor (leading) probe would be "Why do you say 'drugs'? Is it because people on drugs commit so many crimes?" A better probe would be "Can you explain a little more fully what you mean?"

The second problem type has to do with questions asked by respondents. If the question is off the subject of the interview—for example, the respondent asks the interviewer how she likes doing interviews or what she thinks about a particular issue—the interviewer simply says that interviewers are not supposed to discuss their personal feelings during an interview. If the question is about the survey question, the interviewer must refrain from replying unless there are specific instructions on what to say. One never knows how a comment, even an offhand one, might affect an answer.

In response to the third problem—the respondent makes a comment about the question's topic—the interviewer should refrain from comment and lead the respondent back into the interview. As noted in an earlier example, when, in response to another crime survey question, the respondent mentioned that her daughter had been mugged, the interviewer should acknowledge hearing what the respondent said but avoid comment. A neutral filler, such as "I see" can serve this purpose.

Exhibit 9.6 provides examples of how to handle common interview problems. The same type of role-playing exercises can be used to practice these as well, mixing up the respondent behaviors, so each interviewer has to think and react quickly. Remember that in all of these situations, a quick—and correct—reaction may mean the difference between an interview and a break-off and refusal; between reliable, unbiased data and bad data.

Exhibit 9.6   Conventions for Handling Problematic Respondent Behaviors

Respondent interrupts during the question text with an answer:

- The interviewer should read all of the question. However, if it is part of a series of identically structured items, the interviewer should use her judgment whether to accept the response or finish reading the text.

Respondent interrupts during the response categories with an answer:

- If the question is attitudinal, the interviewer should still read all the categories.
- If the question is factual, the interviewer should accept the response.

*Respondent asks what a word or **concept** means:*

- The standard response is to tell the respondent "Whatever it means to you."
- For particular technical terms, all interviewers may have been provided a uniform definition.
- If the respondent says he or she cannot answer without knowing what is meant by a word or term, the interviewer should code it as a "don't know."

Respondent asks for more information or asks about conditions not in the question:

- The interviewer should say, "If your answer is based on just what I've told you, would you say . . . " and then repeat the response categories.
- If the respondent insists he cannot answer without knowing whether a particular condition applies, or without more information, the interviewer should code it as a "don't know."

As we have seen, the data collection process is fraught with potential sources of error. The most serious are generally unit nonresponse, respondent error, and interviewer effects. The number of these types of error, of course, is mainly a result of how well the various aspects of the study are designed and implemented. Still, experience has shown that the incidence of such errors is also closely associated with the data-collection method. Exhibit 9.7 is a guide to the typical levels of these sources of error for three data-collection methods.

Quality Control

Several routine procedures are used to track the quality of telephone survey implementation. Each is related to error sources we have noted as important in most telephone studies.

Three procedures are typically used to track interviewers' performance: monitoring, callback verification, and data checking. In monitoring, a small

Exhibit 9.7   Typical Levels of Error, by Data Collection Method

	Data Collection Method		
	Telephone	Mail	Group Administration
Unit nonresponse	Low	High	Low
Interviewer effects	Medium	N/A	N/A
Respondent errors	Low	Medium–High	Medium–High

sample of each interviewer's cases are listened to during the interview, without the knowledge of either the interviewer or the respondent. This procedure, which requires a central telephone facility with appropriate equipment, enables a trained supervisor to evaluate interviewer behaviors on both a general level (e.g., professional interaction with respondents) and on a very specific level (e.g., verbatim reading of all the questions, proper probing of responses when necessary, and proper handling of respondents' questions or difficulties). The percentage of calls monitored depends on available staff time but should be approximately 10% to 20%. More frequent monitoring is advisable at the outset of the study to try to identify any problems as early as possible for new or less-experienced interviewers and for any interviewers who have had difficulty in either gaining respondents' cooperation or conducting the interview. A key aspect of monitoring is immediate feedback—both positive and negative—to interviewers.

Callback verification is sometimes used if monitoring facilities are not available. In this procedure, a sample of each interviewer's cases are recontacted to verify that the interview was done, to check its length, and to ask whether the respondent had any problems or questions and, if so, how the interviewer reacted to them. There is no set rule about how much verification should be done, but a 10% check is frequently used. Clearly, monitoring is more effective as a quality control procedure and should be considered essential to the conduct of a telephone survey.

For in-person surveys, real-time monitoring cannot be done. Some amount of callback verification is essential. Although interviewer falsification is not a major problem, it does sometimes happen. The temptation to make up an interview (called "curbstoning") is much greater in the field than in a central telephone facility. CAPI systems, which can record interview length, as well as the day and time of the interview, provide a check and make falsification more difficult. The pay schedule can also affect the tendency to falsify interviews. If an interviewer is paid per completed interview, there is more incentive to falsify than if the interviewer is paid by the hour, regardless of the number of interviews the interviewer completes.

Data checking is a procedure in which the collected data are tabulated and inspected during the data-collection period. Data checks can uncover problems in questionnaire administration by interviewers, as well as logic and question-design errors not found earlier. In data checking, frequency distributions are produced for each closed variable, and the verbatim responses to any open-ended items are also generated. These data are examined for such things as skip-pattern errors, patterns of missing items or excessive numbers of missing items, proper use of "other specify" categories (i.e., a check that answers are not being put under "other specify" if they should have been coded into one of the closed categories), consistency between related items, and the clarity of recorded responses to open-ended questions. Data checking is very valuable in spotting problems early enough in the data collection so that, if necessary, corrective actions can be taken.

Self-Administered Surveys: Mail and Internet

Although there is great interest and activity in Internet surveys, whether the data are collected by e-mail or on a Web site, we are still in the early stages of learning how to conduct them with the rigor and quality we expect in other data collection modes. This state of affairs, in itself, suggests caution in deciding when to use Internet data collection and in selecting survey procedures. A discussion of research on Internet surveys is beyond the scope of this book. Suffice it to note that much current practice follows—whether wisely or not, we cannot yet say—the model of conventional mail surveys. One of the main sources on Internet data collection is by Dillman (2000), whose procedures for mail survey design (Dillman, 1978) have been the standard for more than two decades.

A review of issues and perspectives on Internet surveying can be found in the work of Mick Couper (e.g., 2000). In the following guide to mail and Internet data collection, we will use conventional mail procedures as a foundation and note where Internet practices usually differ. This conservative approach is recommended for the novice researcher.

E-mail Versus Web Data Collection

Internet surveys can be conducted by e-mail or on Web sites. In the early days of Internet surveys, data were often collected via e-mail. The questionnaire was either embedded in an e-mail message or attached to it. The respondent answered the questions and returned the e-mail or attachment.

It is still possible to collect data in this manner, but it is not common. We mention it because its modest cost and low reliance on technical skills may, in some instances, make it an attractive alternative.

Three problems led researchers to favor Web site data collection. The wide variety of e-mail systems and of settings within a single system, made it difficult to design a questionnaire that could pass through all systems. For example, many system administrators set size limits on incoming e-mail messages. In those instances, a questionnaire may not get through at all, may be turned into an attachment (which some respondents may be less familiar with handling) or even truncated. Many people have e-mail from one address forwarded to another. If they answer the questionnaire from this second address, the return will not bear the e-mail address the researcher mailed to, making sample tracking difficult.

The second issue is that the available options for designing a questionnaire are typically much more limited in e-mail. If the questionnaire is embedded as an e-mail message, it will simply be text; the responses have to be entered between specified brackets. This increases the chances of respondent error. Skip patterns cannot be handled automatically. Attachments can be executable files that get around some of these problems, but they may be large.

The third issue also has to do with attachments. Because many computer viruses are spread by attachments, many respondents are wary of opening any attachment sent from some person or organization they do not know. This reluctance, of course, will depress response rates. Still, for small surveys with some populations (e.g. students at a university or some membership group) an e-mail survey may be cheap and efficient. We have pointed out some ways that e-mail data collection can contribute to survey error; can you think of some others?

Web surveys are far more common. For this reason, the following discussion about Internet data collection is limited to Web surveys. In these surveys, e-mail is used to contact the respondents, but a URL embedded in the e-mail takes the respondent to a Web site to complete the questionnaire. The Web survey discussion does not mention particular software. Just as with CATI systems, several alternative systems are available. Like other software, the systems change over time or disappear from the market. And just like other software, one has to be wary of bugs, concerned about support, and careful that it is compatible with any other systems it must interact with.

Unit Response

Unit nonresponse is the principal source of nonsampling error in mail surveys, which usually achieve lower response rates than either interviewer-administered general population surveys or surveys of many special populations. Response rates are typically even lower for Web surveys. For both mail and Web surveys intense followup contacts are essential to obtaining acceptable response rates.

One reason for low response rates in these administration modes is that such surveys are simply easier to decline. An interviewer automatically brings some small amount of social pressure to at least listen to the appeal to participate. The interviewer can often tailor her appeals to specific respondent reservations. Surveys that don't use interviewers must rely totally on written or graphic material to convince respondents to participate and then provide instructions to complete the questionnaire. It is difficult to know beforehand how well the materials we design will work. Unlike interviewer-administered surveys, conventional pretesting will not provide much information about how the letter and questionnaire influence response. Poor materials will get a low response rate, which is feedback of a sort, but the conventional pretest will provide little information about what to change. As we have noted, the use of focus groups and cognitive methods, in conjunction with conventional testing, will probably be more helpful.

The problem of nonresponse can be greatly reduced if such surveys are limited to special populations, such as members of an organization, employees in an industry or workplace, students at a university, or some other relatively homogeneous group.

Unlike topics in telephone surveys, the topic of a mail or Internet survey can be a major factor in the decision to respond or not. The respondent is told the topic in the cover letter, can guess it from the envelope's return address, and, of course, can flip through the questionnaire. Usually surveys of special populations are done because the topic particularly applies to them; in those cases, making the topic prominent will generally be an advantage.

The option to preview the questionnaire is usually not possible in Web surveys; whether that is an advantage or a detriment depends on the questionnaire. If the instrument is relatively short and appears easy to complete, respondents would notice this and might be more likely to participate. In the absence of the option to preview the questionnaire, it is important to give some indication of its length in the cover letter.

Prior Notification About the Survey

Prior mail or e-mail notification should be given to respondents informing them that a survey is going to arrive soon. This has become a common practice in conventional mail surveys (Dillman, 2000). It can be a useful way to identify bad addresses (via returns by the post office or e-mails that bounce back as undeliverable). If you have reason to believe that your sampling frame may be somewhat out of date or for other reasons may contain errors, a prenotification can provide an estimate (not a perfect measure) of frame error. This can tell you how much additional sample you need to send in order to reach the target number of respondents in your questionnaire mailing. Also, if the questionnaire is expensive to print and mail, you will save money by eliminating mailing to some bad addresses.

A prior e-mail notification should be sent in Web surveys. Doing so can identify some bad addresses. A more important purpose is its use as an additional opportunity to let respondents know that a legitimate survey is going to be sent to them. Because of the proliferation of spam and other unwanted e-mails, this is no small issue.

Followup Contacts

After the prenotification, how many additional attempts to obtain an interview are made will depend on available resources. It is essential both in mail and Web surveys that *some* followups be sent, even if the number we recommend is not possible. The final effort should use an alternate approach, as described below, which can be very effective in boosting response.

In conventional mail surveys, we follow Dillman's current recommendation of four contacts by first class mail and an additional special contact (Dillman, 2000). These include

- Prenotification letter
- First questionnaire mailing
- Postcard
- Second questionnaire mailing
- Third questionnaire mailing by special means

This same pattern can be used for Web surveys. Of course, in Web surveys, none of the mailings include a questionnaire. The questionnaire always remains posted on the survey Web site. However, in a conventional mail survey, each mailing (except the reminder postcard) should include a copy of the questionnaire; primary reasons for nonreturns is simple misplacement, loss, or (perhaps inadvertent) discarding of the questionnaire (Dillman, 1978, 2000).

In both mail and Web surveys, mailings should be spaced optimally over the data-collection period. The spacing of the followup mailings depends on the flow of returns from the prior mailing. Until returns dwindle to a trickle, there is little advantage (or economic sense) in sending out additional mailings. Although the response pattern will differ for Web surveys, where the response is much more rapid, the logic of waiting for returns to decline is the

Exhibit 9.8   Return Rate in a Mail Survey

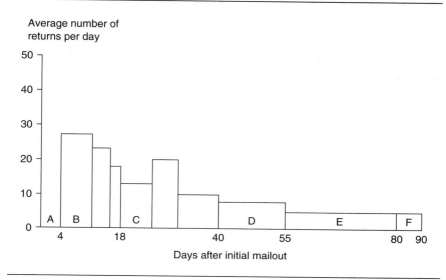

A. initial mailing (day 1)
B. postcard reminder (day 4)
C. second mailing (day 18)

D. third mailing (day 40)
E. reminder calls (days 55–80)
F. special mailing (day 80)

same. Consequently, we must keep track not only of how many conventional returns we receive but also of how many we receive on each day after the initial mailing and each followup mailing.

Response Patterns

Typically, regardless of the survey population, a few people will respond right away to a mail survey; this is followed by a period of relatively high returns, with about the same number each day. The reminder postcard will help to maintain this flow for a while, but then it will decrease. At some point during this decline, we should send out the second mailing, including a questionnaire. To illustrate this typical pattern, Exhibit 9.8 shows the return rate for a national survey of county officials on the subject of how county priorities are determined. In a Web survey, the response pattern is much more clustered toward the beginning of each mailing; those who intend to respond typically do it almost immediately.

During a mail survey, it is useful to track the number of returns per day (or the percentage of the target sample received) as a way to spot response-rate problems. Exhibit 9.8 shows that after the initial mailing there is a lag before returns begin to come in. During this waiting period, a postcard reminder is sent to all sample members. There is a large first-wave response, which trails off after a couple of weeks. At that point the second mailing, with a questionnaire, is sent out, with a similar result. Later phone followups to nonrespondents bring in a few more cases, which finally dwindle to the point that we end data collection.

If, after tracking the initial mailing, we conclude that the rate of return is not what we anticipated, what should we do? First, we should determine whether there are any patterns to the nonresponse. If some sample subgroups are less willing to cooperate than others, we may want to shift some resources to them. This may be done by planning an extra mailing to them or allocating them a larger share of the nonrespondent sample to whom we send a special mailing. If monetary incentives are being used, some increased payment to such groups may help (Warriner, Goyder, Gjertsen, Hohner, & McSpurren, 1996).

Reasons for Nonresponse

Second, it is almost always useful to try to determine the reasons for nonresponse. If telephone numbers are available, a call to some nonrespondents to determine their reasons for not returning the questionnaires may help. If we find in the survey of county officers, for example, that the elected county board members are often not receiving the questionnaire promptly because many are only part-time on the board and spend most of their time at another job, we may want to channel some resources toward telephone calls to obtain these other addresses. Our second mailing would be redirected to the alternate addresses. Remember that in this case, it is the person, not the address, that is the sampling unit.

We might consider these kinds of issues on two other studies. First, we consider a regional survey of recreational boat owners about the problems of trash in public waterways; and second, a study of academic survey organizations about methods they use to develop questionnaires. If we should find that many of the recreational boat owners are concerned about the confidentiality of their answers to questions about trash disposal from their own boats (because they know some disposal practices are illegal), we may want to tailor the cover letter for the second mailing to address this issue. For the survey of survey organizations, we might find that there are delays because some requested information (e.g., the number of surveys that used particular pretest methods) is not readily available. A subsequent cover letter might acknowledge this likely problem but base its appeal on the need for the study to have a good response rate because the results will be reported at an industry conference. The organization's respondents, having often been in the same position themselves, may be motivated by this appeal to put in the extra required effort.

The options open to us during a study are limited, but sometimes a small adjustment can have enough of an effect to boost overall response rates or those of particular subgroups to the extent that we do not have to abandon plans for their separate analysis. The key point, with this and other procedural issues, is that if we see that things are not going well, we take what steps are available to us to improve the situation. Things sometimes turn around on their own, but a good researcher does not count on it.

The information to be tracked in a Web survey is the same. Of course, we are dealing with questionnaires completed on-site (not returns), and the technical methods for tracking the count of completed questionnaires is different. The main difference in Web survey tracking is that the tracking system needs to be checked as part of the pretest. It is a mistake to assume that a system—whether "out of the box" or designed specifically for your study—will work exactly as you expect it to work. When main data collection is underway is not the time to discover problems.

Respondent Selection

Random respondent selection (within some sample unit) is rarely used in mail or Web surveys. These modes are not recommended for general population surveys, so selection within a household is not an issue. Most often our frame will be a list with names of eligible respondents. However, in some business or organization surveys, we may have only a title. Even worse, we may have to ask that our questionnaire go to the (unnamed) person who performs a certain function in the organization (e.g., sets personnel policies), has a certain title (e.g., chief financial officer), or has particular knowledge about the organization (e.g., knows about expenditures for various purposes). The farther away from naming a specific person we are, the greater the likelihood the questionnaire will not reach the correct person and the more difficult it is to obtain a good response rate. Imagine that you are designing a survey of a particular type of business or organization. Think about who would be the person you want to complete your questionnaire and how you would identify or locate that person in the survey.

We lose a lot of control in mail and Web surveys, compared to interviewer-administered studies. For mail and Web surveys, we do not know if an intermediary is the first person to see the mailing or not. The likelihood of this depends heavily on the population surveyed. In business or organization surveys, conventional mail may well be screened before reaching the target respondent; the higher that person is in the organization, the greater the likelihood this will occur. If an alternate means of contact (e.g., telephone) is affordable, it can be very useful in identifying and persuading intermediaries (i.e., gatekeepers) to pass the questionnaire on to the respondent.

We depend on someone to read and follow the instructions about who should complete the questionnaire. Even if the instructions are well written and are followed, in many instances (especially in surveys of very large organizations), the respondent may delegate the completion of the questionnaire to someone else. What types of response error this leads to depends on the study.

Refusal Conversion

Unfortunately, in mail or Web surveys, the range of tactics used in telephone and in-person surveys is not available. Rarely does a respondent send back a mailed questionnaire saying he refuses, let alone why he refuses. So we cannot separate conscious decisions not to participate from nonreturns for other reasons. Thus, it is difficult if not impossible to tailor our followup efforts to the reasons for nonresponse. We are reduced to using general and multiple appeals. For example, a cover letter accompanying a followup mailing might mention things like the respondents' busy schedules or the possible misplacing of the first questionnaire (or having problems accessing the Web site). But the reason for the nonreturn may have nothing to do with either. One method we can make use of on mail studies is the special delivery mailing and/or request for address correction. Both may be effective. The request for address correction should be done early to help ensure that subsequent mailings are sent to the correct address. Special delivery or other special methods, because of their cost, have to be focused on fewer cases later in the study. When using these types of mailings, it is important to keep in mind that the mail should be deliverable whether or not the respondent is at home. Having to go to the post office or make other special efforts to get the mailing are more likely to increase sample members' resentment than to increase response rates. (See Dillman, 1978, for a treatment of this issue in general population mail surveys.)

Samples of Nonrespondents

Finally, we want to consider samples of nonrespondents in situations in which unit nonresponse is high. Samples of nonrespondents are routinely used in mail surveys to assess differences between respondents and nonrespondents, as well as to increase the overall response rate. The method involves selecting a small number of sample members who did not return the mail questionnaire and trying to obtain the interview by another means, usually by telephone. We then compare their characteristics to those of the mail respondents to assess the direction and magnitude of possible bias in the mail sample. While a description of statistical adjustments based on such results is beyond the scope of this book, it is useful simply to know that these "reluctant respondents" are more often of a particular gender or background than the others, or that they tend to be more or less in favor of some proposal referred to in the questionnaire. Nonrespondent sampling is clearly a tactic that has to be planned in advance, as part of the overall design of the survey. But exactly how it is used procedurally can be determined after more is known about the nature of the nonresponse.

Item Nonresponse

Handling missing or ambiguous data can be a major issue in mail surveys. In Web surveys, it is possible to program the questionnaire so that the respondent cannot move forward if the respondent skips a question. This may be a mixed blessing. On the one hand, it certainly reduces item nonresponse; on the other hand, a respondent who feels strongly about skipping a question has the choice of entering false data or refusing to finish the questionnaire. An ethical issue also arises. We typically tell respondents they can skip any question they don't want to answer. We need to think carefully about whether we want to undercut this pledge simply because technology makes it possible.

When a mail questionnaire is received with unanswered items, there are three choices: ignore the items and code as "missing" in the data set; try to determine what the answer "should be" and insert that value into the data set; or recontact respondents to obtain the missing answers.

In making decisions about which option to choose, we must have a sense of which variables are crucial to our study. Obviously, all questions are valuable or we would not have included them, but there are priorities. For example, if the main analysis concerns racial differences, then missing race makes the case much less useful to the study. On the other hand, if differences by race are not central to the study purpose, we would be much less concerned if this item is missing.

If the amount of missing data is both very small (say, less than 3%) and the items are not crucial variables, we want to select the options that use fewest, or no, resources. Either ignore it or see whether the answer can be determined from some combination of other answers in the questionnaire. For example, if gender is missing but elsewhere in the questionnaire the respondent reported having attended a Catholic high school known to be an all-girl institution, we can then confidently enter "female" for the missing item. Great care must be taken in using this process. Errors in such imputation can make the data worse, not better.

When it appears that data are missing because of a misunderstood "skip" instruction and the item is a key variable, it may make sense to recontact the respondent to obtain the information. This is especially appealing if the recontact can be done quickly by telephone. If the calls are long-distance, this cost must be factored in. It is useful to have a small amount of resources—funds and schedule time—set aside for such contingencies.

Tracking

The procedures for tracking mail surveys are fairly straightforward (see Mangione, 1995; Dillman, 2000). We have noted the necessity for tracking mail returns by date of return in order to plan future followup efforts. This sort of tracking can be handled simply with a spreadsheet and a simple system of case id numbers. For Web surveys, tracking can be more complex, but depends greatly on the software application being used. Although consideration of alternative Web survey software is beyond the scope of this book, we do caution the first-time Web researcher to determine the capabilities and ease of use of software for both questionnaire administration and for tracking.

Notes

1. The most comprehensive, though somewhat dated, treatment of survey errors is R. Groves's *Survey Errors and Survey Costs* (1989).

2. Note that the term *survey error* is not particularly directed at mistakes per se, such as incorrectly keying in data from a mail questionnaire, but is broad enough to include these as well.

3. In itself, the process of taking a measurement is also subject to error. Such measurement errors are, of course, not restricted to surveys but are found throughout the empirical sciences.

4. Note that this has nothing to do with intending to elicit a false report, that is, to induce people to report their age as younger or to answer some other question in a certain way. As opposed to everyday usage, bias in survey research indicates effect, not intent.

5. This is one reason we standardly collect demographic information as part of the survey. It allows us to compare our sample's demographic distribution to that of all people in the survey area meeting the target population definition.

6. There are weighting adjustments that can be made after the data are collected to address this problem to some extent. However, as we will see in Chapter 10, weights bring to the survey analysis their own complications and increases in other error sources.

7. Some respondents who refuse to be interviewed are more adamant than others. The less adamant refusals are termed *soft*.

8. In fact, though only recently introduced, the term has already been extended to include all aspects of survey computerization, such as data transmittal, processing, analysis, and dissemination.

9. In a survey conducted as part of a research class, this would mean moving those individuals to coding, monitoring, or other noninterview work.

10. The response rate is defined as interviews divided by eligible households. The cooperation rate is interviews divided by the sum of interviews, partial interviews, and refusals. At the end of the study, response rate is the main measure of data collection success. During data collection, the cooperation rate is a better indicator of how well interviewing is going. Why is this?

11. This is actually the rule, rather than the exception, for surveys of organizations. In planning that type of survey, allowance should be made for one or two calls simply to set up the appointment with the target respondent.

12. For general population surveys, approximately 80% to 90% of calls should be made between about 6 p.m. and 9 p.m. on weekdays, Saturday and Sunday late mornings and afternoons, and Sunday evenings. Calls at other times are mainly to screen out business numbers and to reach respondents with unusual schedules.

13. Reverse directories are telephone directories that are arranged in numerical order by telephone number, rather than alphabetically. Haines and R. H. Donnelley are the main publishers of these directories.

14. In a study comparing respondents from converted refusals to others from several surveys, Blair and Chun (1992) found that there were some small differences in the number of "don't know" answers and item nonresponses between initial cooperators and those who were converted from refusals.

15. Once a random respondent has been selected in a household, that person remains the target respondent for the survey regardless of what happens on subsequent call attempts.

16. One exception can occur when respondents simply overlook an item in a mail survey. For example, in questions with the instruction Mark All That Apply, respondents select fewer items than when asked to answer the same question Yes or No for each item in the series, in effect producing more item nonresponse (Rasinski, Mingay, & Bradburn, 1994).

17. One should also be aware that these interchanges between interviewer and respondent may potentially affect subsequent respondent behaviors (see Couper, 1997).

18. If this type of training is to be conducted properly, a speakerphone should be placed in one or both rooms.

19. The University of Michigan Survey Research Center's *General Interviewing Techniques* (Guensel, Berckmans, & Cannell, 1983) is an excellent source.

20. This should preferably be a 1-800 number if the survey is not local. In any case, it should be a number that is staffed during regular working hours.

21. It is important that surveys in general, and particularly those that might on their surface sound forbidding, not begin with knowledge questions. Remember that all respondents, but especially reluctant ones, still have the option to break off if the first few questions are difficult or make them uneasy.

10

Special Topics

This chapter covers some topics that are not specifically part of survey design and implementation but are nevertheless important. Throughout this book, we have emphasized the importance of resource allocation. Knowing something of how surveys are budgeted is an essential part of understanding use of resources. Data collection involves interaction with potential respondents whom we ask to donate time and resources to our research. There are ethical issues involved in the proper treatment of research subjects. Finally, a record of the survey methodology is considered a basic requirement of research documentation.

Ethical Issues in Survey Research

In the previous chapters, our focus was steadfastly on the obligation we voluntarily assumed to conduct the best research possible within our means. But the responsibility of the serious researcher does not end with conducting the technical and theoretical aspects of the work. We also have obligations to the respondents who agree to participate in our study, to our fellow researchers, and to the users of our results. While an in-depth treatment of the origin and philosophy underlying these issues is beyond the scope of this book, it is essential that the beginning researcher at least be aware of them, if only on a general level.

Two concepts are central to our treatment of respondents: informed consent and protection of confidentiality. While we apply extensive efforts to obtain respondents' cooperation in the survey, the respondents' agreement must be reasonably informed. This means that we must not mislead respondents as to the nature and purpose of the research. We must honestly answer their questions about the project, including who is sponsoring it, its major purposes, the amount of time and effort that will be required of respondents, the general nature of the subject matter, and the use that will be made of the data. We must not badger or try to intimidate respondents either into participating or into answering particular questions after they agree to be interviewed.

Once respondents have agreed to be interviewed, we then assume an obligation to protect the confidentiality of their answers. This is true whether or not we have explicitly told respondents we will do so. Results or data sets that permit the identification of individual respondents should never be made available to others.

These ethical guidelines are recognized by the major professional organizations of survey researchers and are typically overseen by human subjects review committees at universities and other organizations that engage in population research.

These obligations are no less applicable when a project is conducted by a class or a large team of researchers than when a single researcher is involved. In fact, additional cautions may need to be observed in the former situation because there are additional opportunities for inadvertent breaches of these ethical guidelines when many people are privy to the sample and the data.

Revealing or discussing an individual respondent's answers outside of the research group is inappropriate. Also, it is not proper to recontact survey respondents for purposes not related to the research for which they originally agreed to participate. The sample list used for a survey should not be made available to others (even other legitimate researchers) without the additional consent of the respondents. If the data are made available to another party, all identifiers that would permit linking answers to individuals should be removed.

As researchers, we make reasonable efforts to ensure the reliability and validity of our results both through the methods we employ and in our careful interpretation of the resultant data. It is incumbent on us to describe and disclose these methods in our reports and discussions of the survey. There are other reasons, as well, for a full and clear report of the survey methodology. This issue is taken up in detail in the next section.

The Methodology Report

In reporting what methods were used to design and conduct the survey, several technical issues of instrument design, sampling, and data collection are seen in a different light. In the methodology report, the error properties of each survey component are information that speaks to the reliability and validity of the survey's results. Several topics that have been covered earlier are raised again in the methodology report , but more from a user's perspective.

We sought to maximize the reliability and validity of the survey results by means of a series of design and implementation decisions. To assess the soundness of the survey findings, a reader (or user of the data set) must know what key decisions were made and something about the outcomes of the decisions. In their rush to analyze the substantive findings, researchers often omit this final phase of the survey or prepare a report that is incomplete. In this section, we show, through examples, that certain features of the survey methodology can have a crucial effect on the study's findings and their interpretation, and we discuss the key components of a typical methods report.

The Utility of the Methodology Report

The importance of describing the survey's methodological procedures and outcomes is difficult to formalize but can be illustrated quite simply. Consider the following hypothetical finding from the crime survey:

```
Within the past year, have you purchased a gun or other
weapon for protection?
```

Response	distribution
NO	66%
YES	23%
DK/Refused	11%

If it was simply reported that 23% of respondents answered "yes," we would have only part of the picture. How would our confidence in this sample estimate be affected if we knew the study had a 45% response rate? A 78% response rate? Certainly, we would intuitively trust the higher rate to lend more support to the finding. But beyond intuition, the higher response rate is important because we would be less concerned that the non-respondents, had they been interviewed, could have changed the result. Remember that we have a statistical basis for projecting from the *selected* probability sample to the target population. However, if we do not interview all the selected sample members, the *achieved* sample differs from the *selected* sample, and our ability to project to the population is weakened. Because surveys rarely, and only in special circumstances, have a 100% response rate, the potential effect of nonresponse is virtually always an issue. So we have to ask ourselves how the nonrespondents might differ from the respondents.

Of course, we do not know for sure, but our concern is proportional to the magnitude of the nonresponse.[1] If the nonresponse rate is low, then those people we were unable to interview would have to be very different from the respondents in order to change the results. Conversely, if we failed to interview a large percentage of the sample, those people not interviewed could easily change the results.

The distribution of sample responses must also be taken into account in assessing the possible effects of nonresponse. If on a yes/no item we obtained a 93% yes/7% no sample estimate, we would be less concerned about the potential effect of nonresponse on our trust in the estimate than if the sample estimate were 55% yes/45% no. In the former case, even if the nonrespondents are very different from the respondents, it remains likely that the large majority of the population is in the "yes" category. This is not so with the close 55%/45% split, where a substantially different distribution among nonrespondents could change the finding.

Consider two more examples of hypothetical outcomes for the same crime question. What if we knew that the sample was 64% women but that the adult population is only 53% female? Or that 11% of respondents refused to answer this item? If women have generally bought weapons for protection more frequently than have men, we reasonably would want to know whether the data were weighted by sex (i.e., adjusted to match the population distribution), because an unweighted estimate would overestimate the proportion of the total population that has purchased a weapon for protection.

Last, on the issue of item refusals, we might expect that respondents who refused to answer the question are more likely to have purchased a weapon than those who responded. This is true for two reasons: First, many people may consider buying a weapon a private and somewhat sensitive matter and, therefore, may not want to answer. Second, respondents who bought a weapon but did not get a permit or follow other legal requirements may be reluctant to admit the purchase. For both these reasons, we might be concerned that the 23% "yes" figure is an underestimate. In the absence of such supplementary information, it is hard to know what to think about the 23% finding. We can now see how a sample statistic in itself can sometimes be quite uninformative outside of its methodological context.

In many instances, despite strenuous efforts, the distribution of our sample's demographic characteristics differs sufficiently from that of the population whose weights are used to make adjustments. This use of weighting is common enough—and has a large enough potential effect on results—to merit more detailed treatment.

Let us assume that of 800 statewide interviews, 560 (70%) are with urban respondents and 240 (30%) are with rural respondents. Assume further that

Exhibit 10.1   A Comparison of Unweighted and Weighted Data

	Unweighted Estimate		
	Urban	Rural	Total
Yes	224 (.4)	209 (.87)	433 (.54)
No	336 (.6)	31 (.13)	367 (.46)
Total	560	240	800
Percent	70%	30%	100%

	Weighted Estimate		
	Urban	Rural	Total
Yes	272 (.4)	104 (.87)	376 (.47)
No	408 (.6)	16 (.13)	424 (.53)
Total	680	120	800
Percent	85%	15%	100%

on a different yes/no question, these urban respondents were split 40% "yes" and 60% "no," whereas rural respondents answered 87% "yes" and 13% "no." The combined results, shown in Exhibit 10.1, give a statewide sample estimate of 54% "yes" and 46% "no."

Assume that we examine recent census data and find that 85% of the state's adult residents live in urban areas. We then have seriously underrepresented that part of the population in our survey and, consequently, overrepresented rural residents. This incorrect representation affects our state finding. We correct for this by applying a weight that adjusts the urban–rural split to match census data. We do this by counting each urban respondent slightly more than once and each rural respondent less than once. The weight, in this example, is simply the ratio of the census percentage to our sample percentage.[2] The weighted result is shown in the bottom half of Exhibit 10.1. We now have a weighted state sample estimate of 47% "yes" and 53% "no," almost the reverse of the unweighted figure!

This extreme, but not wholly unrealistic, example suggests a few important points about weights that anyone writing a methods report (or analyzing survey data) should be aware of. First, weights can have nontrivial effects on the resulting estimates. They are often required by the sample design (as we saw in Chapter 8 on sampling) or, as in this example, to handle differential nonresponse. They are not optional niceties to be used or not according to the researcher's personal preferences, and they must be handled competently and carefully.[3]

Second, weights are effective in addressing nonresponse bias only if the respondents do not differ significantly from the nonrespondents. We applied a weight to urban respondents because they had a lower response rate than did rural residents and so were underrepresented in the sample. That weight is wholly successful only if the urban respondents are similar on substantive items to the urban nonrespondents. The rather large weights used for purposes of illustration added 120 cases to the urban sample; this is, in effect, the 120 interviews we would have obtained had we achieved perfect urban–rural representation. In essence, we are saying that those "missing" respondents would have answered in the same manner as the ones we interviewed; that is, we kept the weighted yes/no distribution the same as the unweighted distribution.

In truth, we do not know how effective the weight is in addressing nonresponse bias. It is likely that it is at least partially effective. But, because of the uncertainty, we should include a careful discussion of differential nonresponse (particularly if it is as serious as in this example) in the methods report.

Last, we note that the weights have an effect on overall estimates only if there are differences between the groups being weighted. If the yes/no distribution for the urban and rural residents did not differ, the weights would have had no effect on the statewide result *for this particular item*. But, again, we must remember that the survey consists of many variables; some may be affected, others may not.

Although weighting is an important technique, it must be used with caution because when we weight a variable, are potentially affecting or adjusting other variables in the data set. That is, if we weight urban-rural, we may also be adjusting sex, income, race, or other variables—to the extent those variables correlate with the weight variable. Race and income may well vary by urban or rural residence. When we weight on urban-rural, we affect those distributions. If we weight urban up, and there are more nonwhites in the urban areas, their proportion in our total sample will now be higher. On the other hand, if the sex ratio does not differ by urban-rural residence the urban-rural weights will not affect the sex distribution in our total sample.

Another reason for providing a detailed methodological report is to aid comparison to other studies. Returning to the weapons question, if, for example, two separate studies ask the same question but obtain different estimates that are greater than can be accounted for by sampling error, one would want to know why. If the Maryland crime survey, based on 800 interviews, found that 23% of respondents had purchased a weapon, that estimate would have a standard error of approximately 2.9% at the 95% confidence level. If the same question, asked of 700 respondents in Virginia, got a 40% "yes" result, it would be subject to a standard error of 3.6%, also at the 95% confidence level. The range, *based on sampling error alone,* for the Maryland result would be 20.1% to 25.9%; and for the Virginia result, 36.4% to 43.6%. Clearly, the highest estimate for Maryland is much lower than the lowest Virginia estimate.[4]

The different findings may, of course, reflect true differences between the states where the studies were done. It could simply be that more Virginia residents purchase weapons than do residents of Maryland. But such dissimilar results may also be a by-product, or artifact, of the survey methodology.

In trying to understand these differences, we would look first at major aspects of the two surveys. If the method of data collection was not the same, that might account for the difference. Respondents may be more willing by mail to admit obtaining a weapon than they would in a telephone or face-to-face interview. Perhaps the position of the item in the questionnaire might have an effect. If one survey asked the item very early, before the survey's legitimacy or rapport with the respondent were solidly established, it might show lower rates of weapons purchase than a survey that asked the question later in the interview. Finally, if the two surveys were conducted by different organizations (or research teams) the difference might be a result of a "house effect"; that is, something about the way interviewers are generally trained to conduct interviews, such as probing "don't know" responses or not, might affect the results. Of course, the different findings might not be a result of any single methodological difference but of some combination of them.

For example, such a combination might result if the question is asked earlier in the Maryland survey, leading to lower reporting. If the Virginia survey were done by mail instead of by phone, as it was in Maryland, this difference could add to the effect. In this instance, both effects are in the same direction—producing higher Virginia reporting. Such multiple effects could also pull in opposite directions, making our assessment even more complex. Finally, if Virginia has less stringent gun-purchase laws than Maryland,[5] more Virginia residents may have actually bought guns, *and* more Maryland respondents who bought guns may have done so illegally and thus have been reluctant to report the purchase. In this case, the different survey results could partly reflect true population differences and partly reflect a differential willingness to report the behavior. Clearly, trying to make comparisons between surveys may introduce enormous complications; but ignoring them may lead to conclusions that the survey results really do not support.

In all these examples, it is clear that the reader or data analyst must know the details of the survey methods in order to use the results appropriately. Therefore, the careful researcher is obliged to provide information about those aspects of the survey methodology that clearly or potentially bear on the survey's quality or on the interpretation of its findings. Some of this

information, such as sampling errors, can be provided qu
other methodological points are typically descriptive. If we
of methods decisions and their consequences, the list is qu
address all of them would usually be prohibitively costly an
Rather than providing all or nothing, we most often settle
methodological items.

What to Include in the Methodology Report

Exhibit 10.2 lists many of the survey characteristi
included in a methodology report, sometimes called an err
ity profile. We will give brief examples of what might be in
ods report for each of these items and then note guidelines
particular survey, which issues are advisable or essential to
are less crucial.

Sample Design

The methods report often begins with a discussion of th
This section provides a framework for the remainder of the
background for later discussion of sampling frames, weightin

Exhibit 10.2   Quality Profile

Sampling:
- sample design
- sampling errors
- frame problems: coverage

Data Collection:
- instrument development: pretesting special methods, cogni
- data collection: interviewer training, field period, callbac
  monitoring (verification)
- response bias: questionnaire problems
- unit nonresponse: response, refusal, and cooperation rates
- special procedures: samples of nonrespondents, refusal con
- nonresponse bias: differential nonresponse by subgroups

Estimation:
- weights and estimation
- item nonresponse: edits and imputation
- data entry and coding: entry verification and entry errors,

---

other issues. It also begins to tell the reader whether the design was efficient in light of the study's primary objectives.

This section should, at a minimum, include the operational definition of the survey's target population and a general description of the main objectives and features of the sampling method; for example, whether stratification, clustering, or multistage selection was employed.

If stratification was used, was it proportional, disproportional, or some combination of the two? If disproportional stratification was used, the researcher should note whether the main reason was to provide sufficient cases for separate subgroup analysis, to compare subgroups, or something else. Within strata, were elements selected with equal probabilities or not? The definitions of the strata should, of course, be provided as well.

Sampling Errors

Sampling error is the most commonly reported measure of a survey's precision. It is often inadvisably reported in lieu of nonsampling errors. Although this imbalance is inappropriate, every methodology report should still address the issue of sampling errors and, if at all possible, provide sampling errors for all or key variables, or provide a generalized table that can be used for many of the study variables.[6] If the sample in our study is a simple random sample (or a systematic random sample), this task is simple. Statistical packages provide simple random sampling errors (often labeled standard errors[7]) routinely in their output. Unfortunately, in many cases, the sample design may be quite different from a simple random sample.[8] As we have seen, designs often involve stratification, clustering, or multiple stages of selection, frequently along with design, nonresponse, and poststratification weights. In these cases, the simple random sampling errors will overstate (sometimes greatly) the precision of the measures.

What alternatives are available? For many designs, sampling errors can be computed by means of textbook mathematical formulas.[9] These calculations, however, are not necessarily simple, even for a moderate-size data set. The formulas need to be programmed and linked to the data.

A more realistic alternative is to use one of the software packages specifically designed for computing sampling errors (or design effects[10]) that take into account the specific sample design and weighting procedures. The most widely used packages (for the IBM PC) for this purpose are SUDAAN (SUrvey DAta ANalysis) from Research Triangle Institute, and WesVar (http://www.westat.com/wesvar/) from the Westat corporation. The details of their implementation are beyond the scope of this book, but written manuals and short course instruction are available.

If even this cost (or level of programming expertise) is beyond our resources, then we may have to settle for a compromise guesstimate based on design effects noted in studies with designs similar to ours. This is a very risky approach, but one that, in many instances, is preferable either to reporting nothing or to reporting erroneous simple random sampling errors.[11]

Frame Problems: Undercoverage

The main sampling frame issue for the methods report is undercoverage, because presumably any other frame problems, such as overcoverage and multiplicity, were handled in the sampling, data collection, or estimation stages. The kinds of statements we can make are often fairly general, but they may still be informative; for example, "The telephone directories were 8 months old. Therefore, there were some omissions of new residents, but the percentage cannot be specified." Such a statement is useful if, on a particular set of variables, we note differences between long-term and newer residents. Even though the reader of our results cannot specify the bias in the estimates, at least something is known of its direction. Similarly, if we had to use a student directory as the frame at mid-academic year, we might note: "The directory did not contain students who enrolled in the spring semester. Based on past data from the university administration, approximately 10% of entering students enroll in the spring." The importance of including such statements depends on two things: the percentage of undercoverage and its likely relationship to our study variables. Admittedly, these are rough guidelines. Ideally, we would like to know both the direction and the magnitude of undercoverage bias. But if only one is known, that is much better than nothing at all.

When writing the methods report, we should not overlook the positive aspects of our study. It may be that we incorporated a full random-digit dialing (RDD) design for which there is no frame undercoverage. Such design strengths should, here and elsewhere in the report, also be included.

Instrument Development: Pretesting and Special Methods

The methods report should briefly note the number and types of pretests conducted and whether this testing was confined to conventional pretesting or also included cognitive tests, behavior coding, focus groups, expert panels, or combinations of these methods.

If even after pretesting we suspect that some problematic questions remain, this should be discussed. Recall the following question from the student survey:

Most of the time you drank alcohol, on the average, how many drinks did you have at one time?

It could well be that, from conducting cognitive pretest interviews, we learned that respondents who drank alcohol on an irregular basis, say mainly at special social events, had difficulty coming up with an average. Even though we recognized this problem and could not completely eliminate it, the question was so central to our analysis that we decided to include it anyway. It is useful for the reader of our results to know that for these occasional drinkers, the results may be subject to somewhat more measurement error than for the respondent who, say, always has one drink with dinner or occasionally goes to "happy hour" on Friday afternoons. For these and other measurement problems, researchers are sometimes reluctant to mention such shortcomings. But as Sudman (1976) noted about discussions of sample-design biases, far from detracting from the value of the survey, these types of details make it much more useful by giving an indication of the study's limitations. Again, how much of this type of information we choose to include depends on our judgment of the potential effect of such flaws on key variables, or on the readers' appreciation of the survey's strengths, weaknesses, and generalizability.

Data Collection: Interviewer Training, Data Collection Procedures, and Quality Control

The methods report should note when the data were collected (e.g., February through April 2004). This is usually trivial, but it can be important if some variables may have seasonal effects (e.g., a survey asking about recreational activities), or if a public event occurred that might have affected some results (e.g., a widely reported heinous crime that occurred during our crime survey data collection could reduce people's willingness to consider alternative, nonjail sentencing).

A quick summary of the interviewer training should also be given, along with any especially important question-by-question instructions interviewers were given for handling particular problems or questions. For example, consider the question:

Suppose a [fill AGE]-year-old is convicted of selling $[fill AMOUNT] worth of cocaine. He has [fill NUMBER] prior convictions. Do you think he should be sent to prison, required to report to a probation officer once a month, required to report daily, or monitored so his whereabouts are known at ALL times?

1. SENT TO PRISON

2. REQUIRED TO REPORT ONCE A MONTH

3. REQUIRED TO REPORT TO A PROBATION OFFICER DAILY

4. MONITORED SO HIS WHEREABOUTS ARE KNOWN AT ALL TIMES?

8. DK

Respondents might ask various questions such as, Do such monitoring devices really exist? or, What would happen if the person missed reporting to the probation officer? If interviewers were instructed, in the former case, to say that such devices do exist and, in the latter, that the probation officer would decide what would happen, this may be worth mentioning in the report. If there were many instances of such special instructions, a report appendix that listed all instructions to interviewers might be advisable.

The number of callbacks and the general rule for how they were scheduled should be mentioned, as well as the percentage of calls monitored or verified. Many surveys aim to monitor 10% to 15% of interviews, but it is our impression that this number usually includes many interviews for which only portions were monitored.

Response Bias: Questionnaire Problems

If we have reason to believe, say from monitoring interviews in progress or from interviewers' comments, that particular questions caused problems for respondents, this should be mentioned as well. It may be, for example, that many respondents expressed confusion or resisted using the closed-response categories provided for a particular question. In most cases, we will have addressed such issues during the development of the questionnaire. But if, even after these efforts, problems appear to remain, it is the researcher's obligation to report them.

Unit Nonresponse: Response, Refusal, and Cooperation Rates

Unlike some other indicators of survey quality, the survey rates are never an optional report item. The issue is not whether to report, but how. The terms *response, refusal, cooperation,* and *completion rates,* among others, are used by different researchers to mean quite different things. Although there have been recommendations for the computation and presentation of these rates (see Council of American Survey Research Organizations, 1982; Groves, 1989; Groves and Couper, 1998; and Hidiroglou, Drew, & Gray, 1993), they have not been universally adopted. The most recent attempt is a publication by the American Association for Public Opinion Research (AAPOR)(2004) that recommends standard definitions and formulas for telephone and face-to-face surveys. We propose a simple set of rates, but more importantly, we urge that the researcher show exactly how the rates for the particular study were computed and what rules govern the sample dispositions that go into those rates.

In the general population telephone survey, the major call results fall into three areas: nonhouseholds (or other ineligibles), households, and household (or other eligibility) status unknown. Nonhouseholds are fairly clear—numbers that are not working or are connected to businesses or other organizations, hotels, group quarters and so forth.[12] Within the household category, we will have interviews, refusals, and various types of noncontacts for identified (nonrefusal) households where the selected respondent, after the requisite number of contact attempts, was never contacted for a variety of reasons that last the length of the study, such as travel, vacation, an illness or injury, or constant use of an answering machine. In addition, some nonresponse will be a result of households where only languages other than English are spoken. If these cases are eligible according to the study definition, then they must be included in the computation of rates. Finally, even after extensive callbacks, there will remain some (relatively small) number of telephone numbers whose household status cannot be determined. To count either all or none of these as eligible would be wrong; the most reasonable compromise is to consider a percentage of them as eligible households. The best guide we have for this is the household percentage for those numbers we were able to determine (See AAPOR, 2004).

Returning to Exhibit 9.3, we note that of the 1,816 initial sample numbers, 702 were found to be nonhouseholds of various types, 1,061 were households, and 53 could not be determined after 20 calls. We recommend, as is common practice, using the 1,061 identified households as the lower bound on the denominator for computing the response rate and the refusal and noncontact rates, as shown in the exhibit.

To compute an upper bound, we note that 1,763 (1,061 + 702) telephone numbers were classified as households or not; of this total, approximately 60% (1,061/1,763) were households. If we apply this same percentage to the 53 undetermined cases, we would add 32 cases to the denominator, thus decreasing the response rate from 78% (824/1,061) to 75.4% (824/{1,061 + 32}). This procedure provides a range of response rates depending on the assumptions made about the telephone numbers whose household status is unknown.

A final rate that is often presented is the cooperation rate. This is computed as the number of interviews with identified eligibles divided by the interviews plus the refusals and partial interviews plus other. The value of this rate is to show how well the field operation did among those households that were actually contacted *and* included an eligible respondent. This rate is especially valuable in two situations. First, during data collection, we want to track the success of the field operation. While interviewing is going on there are a large number of noncontacts, so the response rate is naturally low and not a good indicator of how well things are going. Second, sometimes the data collection period is so short that few callbacks can be made. This might occur for reasons such as wanting to measure public reaction to a news event. Again, the result is a large number of noncontacts. In both situations the cooperation rate will be a better indicator of interviewing success than the response rate.

Special Procedures: Samples of Nonrespondents and Refusal Conversion

Any special procedures that were used as part of the data collection effort, such as refusal conversion or special followups of samples of nonrespondents, should be described. The description should explain the procedure used and how successful (or not) the effort was. Additionally, if a large proportion of the cases (say, 15% or more) resulted from these efforts, it is useful to report whether these respondents differed on key variables (at statistically significant levels[13]) from the other sample respondents.

Nonresponse Bias: Patterns of Nonresponse and the Use of Weights

If unit nonresponse does not appear to be randomly distributed throughout the population, and especially if the pattern correlates with any substantive variables, a discussion should be included. If, in a statewide survey, for example, there is much poorer response in urban than in rural areas, this should be mentioned, even, perhaps especially, if weighting corrections have been made. If weights are used to adjust for nonresponse, unequal selection probabilities, or poststratification, this should be reported, although the weights themselves may not be given in the methods report.

Estimation

In most small-scale surveys, the estimation procedures—that is, how we convert the sample estimates into estimates of the corresponding population parameters—are very simple, even self-evident. When this is not the case, then the mathematical form of the estimators should be included in the report.

Item Nonresponse and Imputation

In most reports little, if anything, is said about item nonresponse. The guideline for inclusion in the report is whether there is anything unusual. For example, any items that exceed, say, a 5% nonresponse should be mentioned. But as is true for unit nonresponse, the distribution of the responses is an indicator of how important the nonresponse is for the findings. As we have noted, items are often not missing at random but follow a pattern, which may affect our analysis.

Imputation is the substitution of constructed values for items that are either not answered or whose answers are inconsistent with other responses in the same interview. These constructed values are most often based on *other* cases in the data set or on information from other variables in the *same* case. This is not a common practice in small-scale surveys, and we do not recommend it for the novice researcher. If the level of item nonresponse is sufficiently high that key analyses are jeopardized (say for particular subgroups), then imputation may be worth consideration. In such a case, the reader should consult Kalton (1981) or Groves, Dillman, Eltinge, and Little (2002) for a discussion of alternative imputation procedures. Our main recommendation is to develop and test procedures to minimize item nonresponse problems beforehand, rather than to try to correct for them afterward.

Data Entry and Coding: Entry Error, Open-Ends

The error rate, or percentage of all key strokes estimated to be incorrect, for data entry is typically not reported unless there is some extraordinary general problem. A particular difficulty in coding certain items should also be reported. If that does occur, it is usually for open-ended items. Such problems and their solutions should be discussed. In many instances, having monitored the pretest interviews closely and coded the pretest cases for practice researchers will avoid having to explain such problems in the methods report.

Costs and Contingencies: Planning for the Unexpected

Now that we have examined both the general design issues and detailed procedures necessary to complete a survey, and we have some sense of what sorts of unexpected problems can arise, we return to a more detailed look at survey costs. We use the term costs in a broad sense to include materials or services (such as directories or telephone calls) that must be purchased, hourly labor costs (such as for telephone interviewers and data entry operators), and investments of time (such as unpaid student researchers), even though no money has to be spent for it. We take as our example the Maryland crime telephone survey.

Budgeting is an inexact science (as anyone who has ever read a news story about government cost overruns knows), yet an essential one. Discovering, in the course of conducting a survey, that a major cost has been overlooked or underestimated can lead to such disasters as having to cut off data collection before obtaining the target sample size or not conducting the planned followups to obtain a good response rate.

One simple way to approach survey budgeting is to list all the steps necessary to conduct the study and, for each one, to note what resources will be needed and whether there is a cost for them or not. Even if a task does not require payment, listing it will help guarantee realistic planning and, possibly, affect how we want to handle certain tasks. For example, the researcher may have planned to personally type all the questionnaire drafts, handling that part of pretesting for free. Listing the estimated hours needed to do this along with all the other "free" tasks, however, may lead to the decision that the researcher's time is best used elsewhere and that paying a secretary or typing service to handle the questionnaire drafts makes more sense.

By using a task/cost chart like the one in Exhibit 10.3, researchers can figure a budget fairly quickly, *if they know what things cost.* Finding exact costs for some items may be difficult or may require estimates (e.g., telephone charges). But following this procedure forces us to see where the uncertainties are in our budget and, perhaps, where contingencies need to be considered to allow some margin for error. It will also force us to assess realistically the time and cost of the survey. A very common error for the novice researcher is to underestimate the level of effort needed to conduct a first-rate survey.

We begin to construct the task list by listing the survey activities by three time periods: tasks that have to be completed prior to data collection; those that occur during data collection; and those done after data collection. A couple of iterations of the list may be necessary before it is complete, and in the course of making these lists and filling in the costs, some changes in resource allocation may occur, as in the example cited earlier of typing the questionnaire drafts.

In terms of the sheer number of tasks to be done, the list quickly shows us that there is a lot of up-front work to be done to prepare for the study. Although most of the work hours are devoted to the telephone interviewing

Exhibit 10.3   Survey Tasks and Cost Projections: Maryland Crime Survey*

Task	Item	Cost
Pre–Data Collection		
Write outline of survey goals.	8 hrs., draft by researcher	free
Write topic outline of	24 hrs., draft by researcher	free
questionnaire.	16 hrs., review by colleague	free
Find example questions.	24 hrs., student assistant	$288
	4 hrs., review by researcher	free
Write new questions.	16 hrs., draft by researcher	free
	16 hrs., comment by student assistants	$192
Type draft questionnaire.	4 hrs., secretary	$38
Obtain sampling frame.	8 hrs., student assistant	$96
	2 hrs., secretary	$19
	purchase of computerized frame	$650
Select pretest sample.	2 hrs., researcher	free
	6 hrs., student assistant	$72
Prepare training materials.	16 hrs., researcher	free
	20 hrs., professional trainer	$360
	10 hrs., secretary	$95
	materials	$50
Train interviewers for pretest.	16 hrs., professional trainer	$288
	160 hrs., (10) interviewers	$1,280
	48 hrs., (3) interviewer supervisors	$480
	2 hrs., researcher	free
Conduct pretest and debriefing.	4 hrs., professional trainer	$72
	40 hrs., (10) interviewers	$320
	12 hrs., (3) supervisors	$120
	4 hrs., researcher	free
Review results.	6 hrs., researcher	free
	16 hrs., student assistant	$192
Revise and photocopy	16 hrs., researcher	free
questionnaire.	(photocopying)	
	24 hrs., student assistant	$288
Select main sample.	8 hrs., student assistant	$96

*Estimates are based on 800 fifteen-minute telephone interviews in an RDD survey.

(Continued)

Exhibit 10.3   Survey Tasks and Cost Projections: Maryland Crime Survey
(Continued)

Task	Item	Cost
Data Collection		
Train interviewers for main study.	16 hrs., professional trainer	$288
	160 hrs., (10) interviewers	$1,280
	48 hrs., (3) interviewer	
	supervisors	$480
	2 hrs., researcher	free
Conduct 800 fifteen-minute	700 hrs., (10) interviewers	$5,600
interviews and monitor	200 hrs., (3) supervisors	$2,000
(5 weeks data collection).	telephone bill	$1,620
Do data checks.	25 hrs., student assistant	$300
Rework refusals.	20 hrs., interviewers	$240
	10 hrs., supervisors	$100
Post–Data Collection		
Enter data into computer file.	200 hrs., data entry	
	operators	$1,300
	20 hrs., supervisor	$200
Complete data analysis runs.	20 hrs., researcher	free
	40 hrs., student assistant	$480
Write report.	20 hrs., researcher	free
	40 hrs., student assistant	$480
	30 hrs., secretary	$285
Total labor hours	2,155	
Total cost		$19,529

itself, many tasks precede that stage. It should also be noted that most of the cost by far for a telephone survey is for labor; only a small amount goes to materials and expenses. While this balance would shift in the less labor-intensive mail survey, the amount of up-front work still results in significant labor costs for that type of study as well. Looked at from this perspective, one can see that while mail surveys are less expensive than telephone studies, the development costs remain substantial.

While the numbers in Exhibit 10.3 are illustrative, they are approximately correct for surveys conducted by university survey organizations. Of course, if more time is "free"—if, for example, students do all the interviewing—the cost (but not the labor) is less. Often, for university surveys conducted in conjunction with a class, the interviewing is divided between the survey organization's staff and the class. In this way, a reasonable sample size can be achieved without filling all the class time with actual interviewing. As an exercise, you might compute what happens to costs under this and other scenarios of "free" versus paid labor.

Most of the costs and time to do tasks could be estimated or developed by talking with someone who has conducted a survey before (always a good idea during planning). One exception is the number of hours for the interviewing itself. One common way to estimate this is shown in the following calculation:

$$\text{Total interview hours} = \frac{\text{Interview length} \times \text{Sample size} \times \text{I.E.R.}}{60}$$

The only mystery here is the interviewer efficiency rate (I.E.R.), an estimate (based on past experience) of how much time is spent in ancillary activities (dialing nonproductive numbers, handling sample, taking breaks, etc.) for each minute of actual interviewing time. In this calculation, an I.E.R. of 3.5 was used, along with the target 800 fifteen-minute interviews to arrive at the 700 interviewing hours. That is, for every minute of actual interviewing, approximately 2.5 minutes of interviewer time are needed for other activities (such as contacting respondents, dealing with refusals, and so on). Because the I.E.R. is so crucial to a proper calculation of interviewer hours, we recommend seeking the advice of an experienced survey researcher.

The point of this detailed treatment of survey procedures to reduce sources of error is to show that while there are myriad details one must attend to in properly conducting a survey, most are both conceptually and operationally simple. The key is careful planning, realistic allocation of resources and time, and a clear sense of priorities.

For Further Study: Suggested Readings

This short book has focused on the decisions and procedures that the novice researcher needs to address in conducting small-scale surveys. It is our hope that such a narrow focus is most appropriate to the many practical problems the first-time survey researcher confronts, but we realize that it may also give a distorted "cookbook" depiction of the field of survey research. That would be a disservice to both the discipline and to the reader with broader intellectual interests.

In recent years, building on the classic works of the field's pioneer researchers and practitioners, every area of survey research has witnessed high levels of methodological research and conceptual development. In questionnaire development, such seminal works of research and application as Payne's *The Art of Asking Questions* (1951), Schuman and Presser's *Questions and Answers in Attitude Surveys* (1981), and Sudman and Bradburn's *Asking Questions* (1982) have been built on, most notably in the application of cognitive psychology to instrument design and data collection method. Beginning with Jabine, Straf, Tanur, and Tourangeau's *Cognitive Aspects of Survey Methodology* (1984), through Tanur's *Questions About Questions* (1992) and a series of books of collected papers from conferences organized by Schwarz, Sudman, and their colleagues,[14] this area has had explosive growth in the last ten years.

Much of the classic early work on sampling theory and methods, such as Kish's *Survey Sampling* (1965) and Cochran's *Sampling Techniques* (1977), along with practical guides like Sudman's *Applied Sampling* (1976), has been reported in books such as Sarndal, Swensson, and Wretman's *Model-Assisted Survey Sampling* (1992). Even a brief list such as this would be remiss if it omitted key works in data collection such as Dillman's *Mail and Telephone Surveys: The Total Design Method* (1978); in measurement error, Turner and Martin's *Surveying Subjective Phenomena* (2 vols.) (1984); in general design issues, Kish's *Statistical Design for Research* (1987); and the wide-ranging Groves' *Survey Errors and Survey Costs* (1989). Finally, we should note that a series of conferences sponsored by the major survey research professional associations has produced several valuable books on particular areas of survey research. These books include Groves et al., *Telephone Survey Methodology* (1988), Biemer, Groves, Lyberg, Mathiowetz, & Sudman, *Measurement Errors in Surveys* (1991), Kasprzyk, Duncan, Kalton, & Singh, *Panel Surveys* (1989), and Lyberg et al., *Survey Measurement and Process Quality* (1997). Finally, there are journal articles and conference proceedings too numerous to even attempt listing.

Since the first edition of this book nearly a decade ago, much new literature has been produced; more than we can even begin to summarize. Because the sources provided in that edition remain as important and applicable as ever, we limit additional recommendations to books that include treatment of new areas and issues: for Web surveys, Dillman, *Mail and Internet Surveys: The Tailored Design Method* (2000), and Couper et al., *Computer Assisted Survey Information Collection* (1998); or new developments in the use of cognitive psychology in understanding survey response, Tourangeau, Rips, and Rasinski, *The Psychology of Survey Response* (2000), Sirken et al., *Cognition and Survey Research* (1999), and Presser et al., eds., *Methods for Testing and Evaluating Survey Questionnaires* (2004); for concerns with falling response rates and survey quality, Groves et al., *Survey Nonresponse* (2002), and Biemer and Lyberg, *Introduction to Survey Quality* (2003).

While a comprehensive reading of these sources would be of interest only to the professional researcher, we strongly recommend them, even for the relative novice, as references to each aspect of the survey process. Although they contain much theory, they are also readable and give practical advice and explanations of fundamental concepts and concerns that can be found nowhere else.

Even the student who devotes a few days to browsing through these works will be rewarded with useful insights from the best in the discipline, as well as with an overview of an exciting research territory that continues to grow into new areas and evolve using new insights and technologies.

Notes

1. Sometimes *estimates* of this difference are possible from samples of nonrespondents. Typically, such samples are interviewed using an alternative data collection method which, while more expensive, achieves a higher response rate. The most common example is a phone survey followup of a sample of nonrespondents to a mail questionnaire.

2. As an exercise, the reader should compute the weights from these percentages and regenerate the second half of Exhibit 10.1.

3. The standard statistical packages, such as SPSS and SAS, allow for inclusion of weights to produce these estimates. Still, it is useful to do several calculations by hand to get a real sense of how weights affect results.

4. As an exercise, the reader should compute these standard errors at the 68%, 90%, and 99% confidence levels. Note that the calculations ignore the finite population correction and do not take a design effect into account. How might these two factors affect the sampling error estimates?

5. In addition, we assume guns are the main weapon people buy for protection.

6. Such a table is constructed by averaging the sampling errors for a wide range of study variables. These averages are then given in a table for splits on percentages of 90/10, 80/20, 70/30, 60/40, and 50/50. The table can then be used to find the *approximate* sampling errors for particular questions whose results can be expressed as percentage dichotomies. As an exercise, the reader should compute sampling errors at the 95% confidence level (assuming simple random sampling) for subsamples of 100, 200, 300, and 400 respondents. See Exhibit 7.1.

7. The standard error is the standard deviation divided by the square root of the sample size. Typically, standard errors corresponding to 90%, 95%, or 99% confidence levels are given in methods reports.

8. It is very important to remember that an equal probability of selection method (epsem sample) is not at all the equivalent of a simple random sample. Many very complex designs may result in equal probabilities of selection, but their sampling errors are quite different from those of a simple random sample.

9. See, for example, Kish (1965).

10. Design effect is the ratio of the variance of a complex design to that of a simple random sample of the same size (Kish, 1965). Once design effects are available, they can be used to convert simple random sampling standard errors to the correct complex ones (see Chapter 7). These converted sampling errors are then used in the manner discussed above to generate a table of sampling errors appropriate to the particular sample design.

11. If the methods description does not discuss this issue, the unsophisticated reader will sometimes assume that simple random sampling errors apply.

12. For a complete definition of the types of dwelling units, see U.S. Bureau of the Census, *1990 Census of Population and Housing: Guide, Part A. Text* (Washington, DC: U.S. Government Printing Office, 1992) or go to http://factfinder.census.gov.

13. Chi-square tests of cross-tabulations would suffice for this comparison.

14. Among these are Hippler, Schwarz, and Sudman (1987); and Schwarz and Sudman (1992, 1994).

Appendix A

University of Maryland Undergraduate Student Survey

Confidential. Do not sign your name.

For each question, unless otherwise instructed, please circle the number for the ONE response which best reflects your opinion.

The Book Center

1. How often have you shopped at the Book Center this semester, that is, since January?

   ```
   1. just once or twice
   2. less than once a week
   3. about once a week
   4. more than once a week
   5. or not at all ===> go to question 10
   ```

2. Have you shopped for *new textbooks* at the Book Center this semester?

   ```
   0. no ===> Go to question 4
   1. yes
   ```

3. How satisfied were you with the *new textbook's* quality, cost, and selection?

	VERY SATISFIED	SOMEWHAT SATISFIED	SOMEWHAT DISSATISFIED	VERY DISSATISFIED
a. Quality	1	2	3	4
b. Cost	1	2	3	4
c. Selection	1	2	3	4

4. Have you shopped for *general supplies* such as pens and paper at the Book Center this semester?

```
0. no ===> Go to question 6
1. yes
```

5. How satisfied were you with the general supplies' quality, cost, and selection?

	VERY SATISFIED	SOMEWHAT SATISFIED	SOMEWHAT DISSATISFIED	VERY DISSATISFIED
a. Quality	1	2	3	4
b. Cost	1	2	3	4
c. Selection	1	2	3	4

6. Have you used the Book Center this semester to *special order* books or other materials for your classes?

```
0. no ===> go to question 9
1. yes
```

7. How satisfied were you with the special order service?

```
1. very satisfied            3. somewhat dissatisfied
2. somewhat satisfied        4. very dissatisfied
            ⇓                            ⇓
```

7a. What was the MAIN reason for your satisfaction?	b. What was the MAIN reason for your dissatisfaction?
1. timeliness	1. timeliness
2. selection	2. selection
3. cost	3. cost
4. service	4. service
5. something else _____ (please specify)	5. something else _____ (please specify)

8. What one thing can the Book Center do that would most improve the special order services?

_____

_____

_____

_____

9. In general how would you rate the Book Center staff's helpfulness, courtesy, and knowledge of products and services?

	GOOD	EXCELLENT	FAIR	POOR	NO EXPERIENCE/ NO OPINION
a. Helpfulness	1	2	3	4	8
b. Courtesy	1	2	3	4	8
c. Knowledge of products & services	1	2	3	4	8

Academic Advising

10. Since you have been at UMCP, on average, how often have you met with an academic advisor?

    1. Never
    2. Less than once a semester
    3. Once a semester
    4. Twice a semester
    5. Three times a semester
    6. More than three times a semester

11. Is your current advisor

    1. Professional advisor (full-time nonfaculty advisor)
    2. Peer (undergraduate student)
    3. Graduate Assistant
    4. Faculty
    5. Secretary
    8. Don't know
    9. No current advisor

12. What type of advisor do you prefer?

    1. Professional advisor (full-time nonfaculty advisor)
    2. Peer (undergraduate student)
    3. Graduate Assistant
    4. Faculty
    5. Secretary
    6. No preference

13. Have you ever had a peer advisor?

    Yes
    No ===> skip to question 14

13a. Do you think that a peer advisor should be at least the same class rank as the student, or does it not matter?

    1. Should be at least the same class rank
    2. Class rank does not matter

14. How important is it to you that your academic advisor provide:

	VERY IMPORTANT	SOMEWHAT IMPORTANT	NOT VERY IMPORTANT	NOT AT ALL IMPORTANT
a. Information on academic requirements for graduation?	1	2	3	4
b. Referrals to other campus services?	1	2	3	4
c. Walk-in advising sessions?	1	2	3	4
d. Help deciding which courses to take each semester?	1	2	3	4
e. Advice on planning your undergraduate program?	1	2	3	4

15. Since you have been at UMCP, how helpful to you has advising been in general?

    1. Very helpful
    2. Somewhat helpful
    3. Not too helpful
    4. Not at all helpful

16. How helpful has advising been in assisting you in selecting elective courses?

    1. Very helpful
    2. Somewhat helpful
    3. Not too helpful
    4. Not at all helpful
    5. No experience

17. How helpful has advising been to you in making decisions about academic goals?

    1. Very helpful
    2. Somewhat helpful
    3. Not too helpful
    4. Not at all helpful
    5. No experience

18. What is your current grade point average (GPA)?

    1. 3.6-4.0
    2. 3.0-3.5
    3. 2.5-2.9
    4. 2.4 or below

19. Compared with other students of your class rank, do you think your grade point average (GPA) is

    1. a great deal above average
    2. somewhat above average
    3. average
    4. somewhat below average
    5. a great deal below average

20. Compared with students of your class rank and *your race,* do you think your grade point average (GPA) is

    1. a great deal above average
    2. somewhat above average
    3. average
    4. somewhat below average
    5. a great deal below average

21. How frequently has each of the following happened to you?
    [If you don't know (DK) or if you're not sure, circle 8.]

	OFTEN	SOMETIMES	RARELY	NEVER	DK
a. I have received unfair grades because of my race.	1	2	3	4	8
b. I have had teachers of my race who expect me to do better than students of other races.	1	2	3	4	8
c. I have had teachers of other races who expect me to do less well than students of their race.	1	2	3	4	8

Treatment of Black Students

This section is about how you think black students are treated by others.

22. How frequently have you observed the following anti-black behaviors on campus?

	OFTEN	SOMETIMES	RARELY	NEVER
a. Racist comments by students	1	2	3	4
b. Exclusion from social events because of race	1	2	3	4
c. Racist comments by faculty	1	2	3	4
d. Discrimination in awards and honors	1	2	3	4

23. What was the racial make-up of the high school from which you graduated?

    1. About equally black and white
    2. Predominantly black
    3. Predominantly white

Alcohol and Drugs

24. What is the university's maximum penalty for the following violations:

	PROBATION	DRUG TESTING	LOSS OF HOUSING	SUSPENSION	EXPULSION	DON'T KNOW
a. Possession or use of illegal drugs	1	2	3	4	5	8
b. The underage possession of alcohol	1	2	3	4	5	8
c. Sale or distribution of illegal drugs	1	2	3	4	5	8
d. Providing alcohol to someone under 21	1	2	3	4	5	8
e. Drinking on campus	1	2	3	4	5	8

25. Overall, how familiar are you with the campus policies concerning . . .

	VERY FAMILIAR	FAMILIAR	NOT VERY FAMILIAR	UNFAMILIAR
a. Possession or use of illegal drugs	1	2	3	4
b. Sale or distribution of illegal drugs	1	2	3	4
c. Drug testing	1	2	3	4
d. The underage possession of alcohol	1	2	3	4
e. Providing alcohol to someone under 21	1	2	3	4
f. Drinking on campus	1	2	3	4

26. Depending on the circumstances of the case, some students found responsible for the possession and/or use of illegal drugs are given the option of random drug testing for a period of two years in lieu of actual suspension from the University.

How fair do you think this policy is? (circle one)

VERY FAIR	FAIR	UNFAIR	VERY UNFAIR	NO OPINION
1	2	3	4	8

27. How do you feel about . . .

	SUPPORT	STRONGLY SUPPORT	OPPOSE	STRONGLY OPPOSE	IT DEPENDS
a. Undercover investigation on campus	1	2	3	4	9
b. Mandatory expulsion	1	2	3	4	9
c. Random police patrols of residence halls	1	2	3	4	9
d. Laws against drinking alcohol during football games	1	2	3	4	9

28. Assuming you wanted to, how easy do you think it would be for you to get the following on campus?

	VERY EASY	EASY	DIFFICULT	VERY DIFFICULT	IMPOSSIBLE
a. Marijuana/Hashish	1	2	3	4	5
b. Cocaine/Crack	1	2	3	4	5
c. LSD (Acid)	1	2	3	4	5
d. Amphetamines (Speed)	1	2	3	4	5
e. Alcohol	1	2	3	4	5

29. How often in the *last 12 months* have you used the following drugs?

	NEVER	1-2 TIMES	3-10 TIMES	MONTHLY	WEEKLY	DAILY
a. Marijuana/Hashish	1	2	3	4	5	6
b. Cocaine/Crack	1	2	3	4	5	6
c. LSD (Acid)	1	2	3	4	5	6
d. Amphetamines (Speed)	1	2	3	4	5	6
e. Alcohol	1	2	3	4	5	6

30a. Most of the time you drank alcohol, on the average, how many drinks did you have at one time? _____

[One drink = a beer, a mixed drink, or a glass of wine]

☐ Have not drunk alcohol

30b. If you never used illegal drugs, or have not used them in the last year, what was your *main reason for NOT using them*?
(Please circle *one*.)

```
1. It's against my beliefs
2. It's against the law
3. Others disapprove
4. Hard to get
5. Concerns about physical health
6. Did not get desired effect
7. Had bad experience with drugs
8. No desire
9. Other: _____
```

31. If you never used alcohol, or have not used it in the last year, what was your *main reason for NOT using it*?
(Please circle *one*.)

```
1. It's against my beliefs
2. It's against the law
3. Others disapprove
4. Hard to get
5. Concerns about physical health
6. Did not get desired effect
7. Had bad experience with alcohol
8. No desire
9. Other: _____
```

Demographic Information

D1. Are you:

```
1. Male
2. Female
```

D2. Are you:

```
1. White, Non-Hispanic
2. Black or African-American, Non-Hispanic
3. Mexican American, Puerto Rican, or other Hispanic
4. Asian American
5. Native American, American Indian
6. Other: _____
```

D3. Are you currently a

1. Freshman
2. Sophomore
3. Junior
4. Senior

D4. How old were you on your last birthday? _____

D5. In what college are you currently enrolled? _____

D6. What type of institution did you attend just before coming to UMCP?

1. High school
2. Transferred from a 2-year college
3. Transferred from a 4-year college
4. Returning after an absence

THANK YOU FOR COMPLETING THIS QUESTIONNAIRE.

ALL YOUR ANSWERS ARE COMPLETELY CONFIDENTIAL.

PLEASE RETURN THE QUESTIONNAIRE IN THE ENCLOSED ENVELOPE.

Appendix B

Maryland Crime Survey

Hello, I'm calling from the University of Maryland. My name is _____.
We are doing a study for the state of Maryland's Summit on Violent Street
Crime. I need to speak with the adult in your household, who is 18 or older
and will have the NEXT birthday. Who would that be?

1. In general, would you say that the crime problem in YOUR NEIGH-
   BORHOOD is very serious, somewhat serious, not very serious, or
   not at all serious?

   1. very serious
   2. somewhat serious
   3. not very serious
   4. not at all serious
   8. dk

2. In general, would you say that the crime problem in THE STATE is
   very serious, somewhat serious, not very serious, or not at all serious?

   1. very serious
   2. somewhat serious
   3. not very serious
   4. not at all serious
   8. dk

3. In the past year, would you say that there is more VIOLENT crime
   in YOUR NEIGHBORHOOD, less, or about the same?

   1. more
   2. less
   3. about the same
   8. dk

4. In the past year, would you say that there was more VIOLENT crime in THE STATE, less, or about the same?

   1. more
   2. less
   3. about the same
   8. dk

5. In the past year, did anything happen to you or your property that you thought was a crime or an attempt at a crime?

   0. no   [go to question 7]
   1. yes
   8. dk

6. Were you personally threatened or was force used against you?

   0. no
   1. yes
   8. dk

7. In the past year, did anything happen to anyone (else) in your household that they thought was a crime or an attempt at a crime?

   0. no   [go to question 9]
   1. yes
   8. dk

8. Were they personally threatened or was force used against them?

   0. no
   1. yes
   8. dk

9. In the coming year, how likely it is that you, or someone in your household, will be robbed or mugged? Would you say:

   1. very likely
   2. somewhat likely
   3. not too likely
   4. not at all likely
   8. dk

10. In the coming year, how likely is it that your home will be broken into? Would you say:

   1. very likely
   2. somewhat likely
   3. not too likely
   4. not at all likely
   8. dk

Next I want to ask you about some things you may have done within the past year to protect yourself or your family against crime.

11. Within the past year, have you joined a community crime-prevention program?

    0. no
    1. yes
    8. dk

12. (Within the past year, have you) had extra locks put on doors or windows?

    0. no
    1. yes
    8. dk

13. (Within the past year, have you) had a burglar alarm installed in your home?

    0. no
    1. yes
    8. dk

14. (Within the past year, have you) taken self-defense training?

    0. no
    1. yes
    8. dk

15. (Within the past year, have you) gotten a dog for protection?

    0. no
    1. yes
    8. dk

16. (Within the past year, have you) purchased a gun or other weapon for protection?

    0. no
    1. yes
    8. dk

17. When you are not at home, how often do you leave your lights on because of concerns about crime? Would you say

    1. always
    2. frequently
    3. seldom
    4. or never
    8. dk

18. When you are out, how often do you have neighbors watch your home because of concerns about crime? Would you say

    1. always
    2. frequently

```
3. seldom
4. or never
8. dk
```

19.   How often do you avoid certain areas that you'd like to go to because of concerns about crime? Would you say

```
1. always
2. frequently
3. seldom
4. or never
8. dk
```

20.   How often do you avoid going out alone because of concerns about crime? Would you say

```
1. always
2. frequently
3. seldom
4. or never
8. dk
```

21.   How often do you carry a weapon for protection because of concerns about crime? Would you say

```
1. always
2. frequently
3. seldom
4. or never
8. dk
```

Three important parts of the criminal justice system are the police, the courts, and the prisons. I'd like to ask you about the job each is doing.

22.   Do you think that the police are doing an excellent, good, fair, or poor job?

```
1. excellent [go to 23]
2. good [go to 23]
3. fair
4. poor
8. dk [go to 23]
```

22a. Why is that? _____

23.   How do you feel the courts are doing? (Are they doing an excellent, good, fair or poor job?)

```
1. excellent [go to 24]
2. good [go to 24]
3. fair
4. poor
8. dk [go to 24]
```

23a. Why is that? _____

24. (How about) the prisons? (Are they doing an excellent, good, fair, or poor job?)

```
1. excellent [go to 25]
2. good [go to 25]
3. fair
4. poor
8. dk [go to 25]
```

24a. Why is that? _____

25. Would you be willing to pay $100 a year in extra taxes to build more prisons in Maryland?

```
0. no
1. yes
8. dk
```

26. Not everyone convicted of a crime can be put in prison. Maryland prisons are seriously overcrowded. It costs $17,000 to keep someone in prison for a year.

As a result, many people are not sent to prison but sentenced to report to a probation officer ONCE A MONTH.

One alternative would be to require reporting on a DAILY basis.

Another alternative would be to monitor individuals so that their whereabouts are known at ALL times.

I'd like to ask you when you think these alternative sentences would be appropriate.

{AGE OF PERSON, VALUE OF COCAINE AND CRIMINAL RECORD RANDOMIZED}

26a. Suppose a [fill AGE]-year-old is convicted of selling $[fill AMOUNT] worth of cocaine. He has [fill NUMBER] prior convictions.

Do you think he should be sent to prison, required to report to a probation officer ONCE A MONTH, required to report DAILY, or monitored so his whereabouts are known at ALL times?

```
1. sent to prison
2. required to report once a month [go to 27a]
3. required to report to a probation officer daily
   [go to 27a]
4. monitored so his whereabouts are known at all times
   [go to 27a]
8. dk
```

26w. Should he serve some prison time and then be released and monitored at ALL times for the rest of his sentence, OR should he spend all of his sentence in prison?

```
1. released and monitored
2. all in prison
8. dk
```

WEAPON ALSO RANDOMIZED

27a. Suppose a [fill AGE]-year-old is convicted of robbing someone of $100 on the street. He threatened the victim with a [fill TYPE OF WEAPON]. He has [fill NUMBER] prior convictions.

Do you think he should be sent to prison, required to report to a probation officer ONCE A MONTH, required to report DAILY, or monitored so his whereabouts are known at ALL times?

```
1. sent to prison
2. required to report once a month [go to 28a]
3. required to report to a probation officer daily
   [go to 28a]
4. monitored so his whereabouts are known at all times?
   [go to 28a]
8. dk
```

27w. Should he serve some prison time and then be released and monitored at ALL times for the rest of his sentence, OR should he spend all of his sentence in prison?

```
1. released and monitored
2. all in prison
8. dk
```

VALUE OF PROPERTY RANDOMIZED

28a. Suppose a [fill AGE]-year-old is convicted of breaking into a home and taking property worth $[fill VALUE]. He has [fill NUMBER] prior convictions.

Do you think he should be sent to prison, required to report to a probation officer ONCE A MONTH, required to report DAILY, or monitored so his whereabouts are known at ALL times?

```
1. sent to prison
2. required to report once a month [go to 29a]
3. required to report to a probation officer daily
   [go to 29a]
4. monitored so his whereabouts are known at all times
   [go to 29a]
8. dk
```

28w. Should he serve some prison time and then be released and monitored at ALL times for the rest of his sentence, OR should he spend
all of his sentence in prison?

```
1. released and monitored
2. all in prison
8. dk
```

SERIOUSNESS OF INJURY VARIES

29a. Suppose a [fill AGE]-year-old is convicted of starting a fight with a
stranger. The victim's injuries did [fill REQUIRE/NOT REQUIRE]
treatment by a doctor. The offender has [fill NUMBER] prior
convictions.

Do you think he should be sent to prison, required to report to a probation officer ONCE A MONTH, required to report DAILY, or
monitored so his whereabouts are known at ALL times?

```
1. sent to prison
2. required to report once a month [go to 30]
3. required to report to a probation officer daily
   [go to 30]
4. monitored so his whereabouts are known at all times
   [go to 30]
8. dk
```

29w. Should he serve some prison time and then be released and
monitored at ALL times for the rest of his sentence, OR should he
spend all of his sentence in prison?

```
1. released and monitored
2. all in prison
8. dk
```

30. What do you think is the MAIN cause of violent street crime in
Maryland?

** RECORD ONLY ONE ANSWER

_____

_____

31. Other than building more prisons, if you could suggest two things to
the governor or the legislature to deal with the crime problem, what
would they be?

```
1. _____
```

```
2. _____
```

D1. Finally, I'd like to ask you some background questions.

Including yourself, how many adults 18 years of age or older live in this household?

```
01-10    record actual number
11       more than 10
99       na-ref
```

D2. How many children younger than 18 live in this household?

```
0-7      record actual number
8        more than 7
99       na-ref
```

D3. In what year were you born?

```
00       before 1900
01-74    19__
88       dk
99       na-ref
```

D4. What is the last grade or year of school you completed?

```
0        none
1-7      some elementary
8        elementary school
9-11     some high school
12       high school grad
13-15    some college
16       college grad
17       some graduate school
18       graduate or professional degree
99       ref
```

D5a. Are you of Spanish or Hispanic origin or descent?

```
0. no
1. yes
```

D5b. Are you:

```
1. White
2. Black
3. Asian [do not ask if yes to D5a]
4. or some other race: (SPECIFY) _____
9. ref
```

D6. Do you own your home or do you rent it?

```
1. own
2. rent
3. other
9. ref
```

D7. Are you currently:

```
1. married
2. separated
3. divorced
4. widowed
5. or have you never been married
9. ref
```

D8. Are you currently:

```
1. employed full time
2. part-time
3. or not employed at all
9. na-ref
```

D9. All together, how many years have you lived at your present address?

```
00      less than one year
01-50   record actual number
51      more than 50 years
88      dk
99      na-ref
```

D10. All together, how many different phone NUMBERS does your household have for nonbusiness use?

```
1-6   record
7     7 or more
8     dk
9     ref
```

D11. If you added together all the yearly incomes, before taxes, of all the members of your household for last year, 1991, would the total be more than $30,000?

```
0. no [go to D11a]
1. yes [go to D11c]
9. ref
```

D11a. Was it more than $20,000?

```
0. no [go to D11b]
1. yes [go to D12]
9. ref
```

D11b. Was it more than $12,000?

```
0. no [go to D12]
1. yes [go to D12]
9. ref
```

D11c. Was it more than $50,000?

```
0. no [go to D12]
1. yes [go to D11d]
9. ref
```

D11d. Was it more than $75,000?

```
0. no [go to D12]
1. yes [go to D11e]
9. ref
```

D11e. Was it more than $100,000?

```
0. no
1. yes
9. ref
```

D12. In what county do you live?

D13. And, your phone number is [fill in SAMPLE TELEPHONE NUMBER]

```
0. no - what number have I reached? (Specify)
1. yes
9. ref
```

Those are all the questions I have. Thank you for your time.

Appendix C

American Association for Public Opinion Research Code of Professional Ethics and Practices*

We, the members of the American Association for Public Opinion Research, subscribe to the principles expressed in the following code. Our goals are to support sound and ethical practice in the conduct of public opinion research and in the use of such research for policy-and decision making in the public and private sectors, as well as to improve public understanding of survey research methods and the proper use of opinion research results.

We pledge ourselves to maintain high standards of scientific competence and integrity in conducting, analyzing, and reporting our work; in our relations with survey respondents; with our clients; with those who eventually use the research for decision-making purposes; and with the general public. We further pledge ourselves to reject all tasks or assignments that would require activities inconsistent with the principles of this code.

*AAPOR members are to vote on this version, Fall 2004.

The Code

I. Principles of Professional Practice in the Conduct of Our Work
   A. We shall exercise due care in developing research designs and survey instruments, and in collecting, processing, and analyzing data, taking all reasonable steps to assure the reliability and validity of results.
      1. We shall recommend and employ only those tools and methods of analysis that, in our professional judgment, are well suited to the research problem at hand.
      2. We shall not select research tools and methods of analysis that yield misleading conclusions.
      3. We shall not knowingly make interpretations of research results that are inconsistent with the data available, nor shall we tacitly permit such interpretations.
      4. We shall not knowingly imply that interpretations should be accorded greater confidence than the data actually warrant.
   B. We shall describe our methods and findings accurately and in appropriate detail in all research reports, adhering to the standards for minimal disclosure specified in Section III.
   C. If any of our work becomes the subject of a formal investigation of an alleged violation of this Code, undertaken with the approval of the AAPOR Executive Council, we shall provide additional information on the survey in such detail that a fellow survey practitioner would be able to conduct a professional evaluation of the survey.

II. Principles of Professional Responsibility in Our Dealings with People
   D. The Public:
      1. When preparing a report for public release we shall ensure that the findings and any interpretations or implications are a balanced and accurate portrayal of the survey results.
      2. If we become aware of the appearance in public of serious inaccuracies regarding our research, that is, descriptions that are incorrect, distorted, or incomplete, we shall publicly disclose what is required to correct these inaccuracies, including, as appropriate, a statement to the public media, legislative body, regulatory agency, or other appropriate group, to which the inaccuracies were presented.
      3. We shall inform those for whom we conduct publicly released surveys that AAPOR standards require members to release minimal information about such surveys, and we shall make every effort to encourage clients to subscribe to our standards for minimal disclosure in their releases.

E. Clients or Sponsors:

   1. When undertaking work for a private client, we shall hold confidential all proprietary information obtained about the client and about the conduct and findings of the research undertaken for the client, except when the dissemination of the information is expressly authorized by the client, or when disclosure becomes necessary under terms of Section I-C or D of this Code.[1]

   2. We shall be mindful of the limitations of our techniques and capabilities and shall accept only those research assignments which we can reasonably expect to accomplish within these limitations.

F. The Profession:

   1. We recognize our responsibility to the science of public opinion research to disseminate as freely as possible the ideas and findings which emerge from our research.

   2. We shall not cite our membership in the Association as evidence of professional competence, since the Association does not so certify any persons or organizations.

G. The Respondent:

   1. We shall avoid practices or methods that may harm, humiliate, or seriously mislead survey respondents.

   2. We shall ensure that respondents are not subjected to any personal harassment and/or unnecessary and unwanted intrusion into their privacy.

   3. With few exceptions, participation in surveys is voluntary. We shall provide to sample persons sufficient description of the survey to permit them to make an informed and free decision about their participation.

   4. We shall not misrepresent our research or conduct other activities (such as sales, fund raising, or political campaigning) under the guise of conducting research.

   5. Unless the respondent waives confidentiality for specified uses, we shall hold as privileged and confidential all information that might identify a respondent with his or her responses. We shall also not disclose or use the names of respondents for non-research purposes unless the respondents grant us permission to do so. If such permission is requested, respondents must be given a sound reason for the request.

---

[1]Author's note: State universities and other public institutions typically can undertake only those studies whose results may be made public after a specified time period.

6. We understand that the use of our survey results in a legal proceeding does not relieve us of our ethical obligation to keep confidential all respondent identifiable information or lessen the importance of respondent anonymity.

III. Standards for Minimal Disclosure

Good professional practice imposes the obligation upon all public opinion researchers to include, in any report of research results, or to make available when that report is released, certain essential information about how the research was conducted. At a minimum, the following items should be disclosed:

1. Who sponsored the survey and who conducted it.
2. The exact wording of questions asked, including the text of any preceding instruction or explanation to the interviewer or respondents that might reasonably be expected to affect the response.
3. A definition of the population under study, and a description of the sampling frame used to identify this population.
4. A description of the sample design, giving a clear indication of the method by which the respondents were selected by the researcher.
5. Size of samples and, where appropriate, eligibility criteria, screening procedures, and response rates computed according to AAPOR Standard Definitions. At a minimum, a summary of sample dispositions should be provided so that response rates could readily be computed.
6. A discussion of the precision of the findings, including, if appropriate, estimates of sampling error, and a description of any weighting or estimating procedures used.
7. Which results are based on parts of the sample, rather than on the total sample, and the size of such parts.
8. Method, location, and dates of data collection.

From time to time, AAPOR Council may issue guidelines and recommendations on best practices with regard to the release, design, and conduct of surveys.

March 2004

References

Aday, L. A. (1996). *Designing and conducting health surveys: A comprehensive guide.* San Francisco: Jossey-Bass.

American Association for Public Opinion Research (AAPOR). (1998). *Standard definitions: Final dispositions of case codes and outcome rates for RDD telephone surveys and in-person household surveys.* Ann Arbor, MI: AAPOR.

American Association for Public Opinion Research (AAPOR). (2004). *Standard definitions: Final dispositions of case codes and outcome rates for surveys* (3rd ed.). Lenexa, KS: AAPOR.

Aneshensel, C. S., Frerichs, R. R., Clark, V. A., & Yokopenic, P. A. (1982). Measuring depression in the community: A comparison of telephone and personal interviews. *Public Opinion Quarterly, 46,* 110–121.

Belinfante, A. (2004). *Telephone penetration by income by state.* (FCC News Release). Washington, DC: Industry Analysis and Technology Division, Wireline Competition Bureau, Federal Communication Commission.

Belson, W. A. (1981). *The design and understanding of survey questions.* Brookfield, VT: Gower.

Biemer, P. P., Groves, R. M., Lyberg, L. E., Mathiowetz, N. A., & Sudman, S. (Eds.). (1991). *Measurement errors in surveys.* New York: John Wiley & Sons.

Biemer, P. P., Lyberg, L. E. (2003). *Introduction to survey quality.* New York: John Wiley & Sons.

Binson, D., Murphy, P. A., & Keer, D. (1987, May). *Threatening questions for the public in a survey about AIDS.* Paper presented at annual meeting of American Association for Public Opinion Research, Hershey, PA.

Bischoping, K. (1989). An evaluation of interviewer debriefing in survey pretests. In C. Cannell, L. Oksenberg, F. Fowler, G. Kalton, & K. Bischoping (Eds.), *New techniques for pretesting survey questions.* (Final Report for Grant HS05616, National Center for Health Services Research). Ann Arbor: University of Michigan, unpublished report.

Blair, J., & Chun, Y. (1992). *Quality of data from converted refusals in telephone surveys.* Paper presented at the American Association for Public Opinion Research Conference, St. Petersburg, FL.

Blair, J., Menon, G., & Bickart, B. (1991). Measurement effects in self vs. proxy response to survey questions: An information-processing perspective. In

P. P. Biemer, R. M. Groves, L. E. Lyberg, N. A. Mathiowetz, & S. Sudman (Eds.), *Measurement errors in surveys* (pp. 145–166). New York: John Wiley & Sons.

Bradburn, N. M., & Miles, C. (1979). Vague quantifiers. *Public Opinion Quarterly, 43*(1), 92–101.

Bradburn, N. M., Sudman, S., & Associates. (1979). *Improving interview method and questionnaire design: Response effects to threatening questions in survey research.* San Francisco: Jossey-Bass.

Brick, J. M., Waksberg, J., Kulp, D., & Starer, A. (1995). Bias in list-assisted telephone samples. *Public Opinion Quarterly, 59*(2), 218–235.

Bryant, B. E. (1975). Respondent selection in a time of changing household composition. *Journal of Marketing Research, 12,* 129–135.

Campanelli, P. C., Rothgeb, J. M., Esposito, J. L., & Polivka, A. E. (1991). *Methodologies for evaluating survey questions: An illustration from a CPS CATI/RDD Test.* Paper presented at the annual meeting of the American Association for Public Opinion Research, Phoenix, AZ.

Campbell, B. A. (1981). Race-of-interviewer effects among southern adolescents. *Public Opinion Quarterly 45,* 231–244.

Cannell, C. F., Fowler, P. J., & Marquis, K. H. (1968). The influence of interviewer and respondent psychological and behavioral variables on the reporting in household interviews. *Vital and Health Statistics,* Series 2, No. 26. Washington, DC: U.S. Government Printing Office.

Cannell, C., Oksenberg, L., Fowler, F., Kalton, G., & Bischoping, K. (1989). *New techniques for pretesting survey questions.* (Final Report for Grant HS05616, National Center for Health Services Research). Ann Arbor: University of Michigan, Unpublished report.

Caron, J. (1992). *An introduction to psycholinguistics* (T. Pownall, Trans.). New York: Harvester Wheatsheaf. (Original work published 1989)

Carroll, R. (2003, August 4). Millions getting rid of landline phones. *Associated Press.*

Casady, R., & Lepkowski, J. (1993). Stratified telephone survey designs. *Survey Methodology, 19*(1), 103–113.

Catania, J. A., Binson, D., Canchola, J., Pollack, L. M., Hauck, W., & Coates, T. J. (1996). Interviewer and question effects on sex items. *Public Opinion Quarterly, 60*(3), 345–375.

Cho, H., & LaRose, R. (1999). Privacy issues in Internet surveys. *Social Science Computer Review, 17,* 421–434.

Church, A. H. (1993). Incentives in mail surveys: A meta-analysis. *Public Opinion Quarterly, 57*(1), 62–79.

Cobanoglu, C., Warde, B., & Moreo, P. J. (2001). A comparison of mail, fax and Web-based survey methods. *International Journal of Market Research, 43*(4), 441–452.

Cochran, W. G. (1977). *Sampling techniques* (3rd ed.). New York: John Wiley & Sons.

Cohen, J. (1960). A coefficient of agreement for nominal scales. *Educational and Psychological Measurement,* Vol. 20, No. 1.

Cohen, J. (1988). *Statistical power analysis for the behavioral sciences* (2nd ed.). Hillsdale, NJ: Lawrence Erlbaum.

Conrad, F., & Blair J. (2003). Aspects of data quality in cognitive interviews: The case of verbal reports. In Presser, S., Rothgeb., J., Couper, M., Lessler, J., Martin, E., Martin, J., et al. (Eds.). *Methods for testing and evaluating survey questionnaires*. New York: John Wiley & Sons. 2004.

Conrad, F., & Blair, J. (2004). Data quality in cognitive interviews: The case of verbal reports. In Presser, S., Rothgeb, J. M., Couper, M. P., Lessler, J. T., Martin, E., Martin, J., et al., (Eds.), *Methods for testing and evaluating survey questionnaires*. New York: John Wiley & Sons.

Conrad, F., Blair J., & Tracy E. (1999). *Verbal reports are data! A theoretical approach to the analysis of cognitive interviews.* Proceedings of the 1999 Federal Committee on Statistical Methodology Research Conference, Statistical Policy, Working Paper #30, Washington, DC, June 2000.

Converse, J. M., & Presser, S. (1986). *Survey questions: Handcrafting the standardized questionnaire.* Newbury Park, CA: Sage.

Council of American Survey Research Organizations (CASRO). (1982). *Report of the CASRO completion rates task force.* New York: Audits and Surveys, Inc., Unpublished report.

Couper, M. P. (1997). Survey introductions and data quality. *Public Opinion Quarterly, 61*(2), 317–338.

Couper, M. P. (2000). Web surveys: A review of issues and approaches. *Public Opinion Quarterly, 64*(4), 464–494.

Couper, M. P. (2001). *Designing effective Web surveys.* Workshop presented at annual meeting of American Association for Public Opinion Research. Montreal, Quebec, Canada.

Couper, M. P., Baker, R. P., Bethlehem, J., Clark, C. Z. F., Martin, J., Nicholls, W. L., et al. (Eds). (1998). *Computer assisted survey information collection.* New York: John Wiley & Sons.

Couper, M. P., Traugott, M. W., & Lamias, M. J. (2001). Web survey design and administration. *Public Opinion Quarterly, 65*(2), 230–253.

Current Population Survey. (2002). *Telephone in the household by poverty and income level by geographic region.* [MRDF] March Supplement 2002. Retrieved February 16, 2004 from http://ferret.bls.census.gov.

Curtin, R., Presser S., & Singer E. (2000). The effects of response rate changes on the index of consumer sentiment. *Public Opinion Quarterly, 64*(4), 413–428.

Czaja, R. F. (1987–1988). Asking sensitive behavioral questions in telephone interviews. *International Quarterly of Community Health Education, 8,* 23–31.

Czaja, R. F., Blair, J., & Sebestik, J. P. (1982). Respondent selection in a telephone survey: A comparison of three techniques. *Journal of Marketing Research 21,* 381–385.

Davis, D. K., Blair, J., Gourdreau, N., Boone, M. S., Johnson, L. R., & Robles, E. (1998). *Research on race and Hispanic origin for Census 2000.* Report to Population Division, Bureau of the Census.

Davis, J. A., & Smith, T. W. (1991). *General social surveys, 1972–1991: Cumulative codebook.* Chicago: National Opinion Research Center.

Davis, J. A., & Smith, T. W. (1993). *General social surveys, 1972–1993: Cumulative codebook.* Chicago: National Opinion Research Center.

Davis, J. A., Smith, T. W., & Marsden, P. V. (2001). *General social surveys codebook.* Principal investigator, James A. Davis; director and co-principal investigator, Tom W. Smith. Chicago: National Opinion Research Center.

de Leeuw, E. (1992). *Data quality in mail, telephone and face-to-face surveys.* Amsterdam: TT-Publikaties.

Demaio, T. J. (1983). *Approaches to developing questions.* (Statistical Policy Working Paper 10). Washington, DC: Statistical Policy Office, U.S. Office of Management and Budget.

Dillman, D. A. (1978). *Mail and telephone surveys: The total design method.* New York: John Wiley & Sons.

Dillman, D.A. (2000). *Mail and Internet surveys. The tailored design method.* New York: John Wiley & Sons.

Ericsson, K. A., & Simon, H. A. (1993). *Protocol analysis: Verbal reports as data, revised edition.* Cambridge, MA: The MIT Press.

Everitt, B., & Hay, D. (1992). *Talking about statistics: A psychologist's guide to design & analysis.* New York: Halsted Press.

Ezzati-Rice, T. M., White, A. A., Mosher, W. D., & Sanchez, M. E. (1995). *Time, dollars, and data: Succeeding with remuneration in health surveys.* Paper presented at the Federal Committee on Statistical Methodology Conference: Seminar on New Directions in Statistical Methodology, Washington DC.

Federal Communication Commission. (2003). Trends in telephone service. In *Telephone penetration and telephone subscribership* (pp 16–17). Washington, DC: Author.

Fleiss, J. L. (1981). *Statistical methods for rates and proportions* (2nd ed.). New York: John Wiley & Sons.

Fowler, F. J. (1984). *Survey research methods.* Newbury Park, CA: Sage.

Fowler, F. J. (1989). Coding behavior in pretests to identify unclear questions. In F. J. Fowler (Ed.), *Conference proceedings, health survey research methods* (pp. 9–12). Washington, DC: National Center for Health Services Research.

Fowler, F. J. (1995). *Improving survey questions: Design and evaluation.* Newbury Park, CA: Sage.

Fowler, F. J., & Mangione, T. W. (1990). *Standardized survey interviewing.* Newbury Park, CA: Sage.

Frey, J. H. (1989). *Survey research by telephone.* Newbury Park, CA: Sage.

Glenberg, A. M. (1996). *Learning from data: An introduction to statistical reasoning* (2nd ed.). Mahwah, NJ: Lawrence Erlbaum.

Greenbaum, T. L. (1998). *The handbook for focus group research* (2nd ed.). Thousand Oaks, CA: Sage.

Groves, R. M. (1987). Research on survey data quality. *Public Opinion Quarterly, 51,* 5156–5172.

Groves, R. M. (1989). *Survey errors and survey costs.* New York: John Wiley & Sons.

Groves, R. M., Biemer, P. P., Lyberg, L. E., Massey, J. T., Nicholls, II, W. L., & Waksberg, J. (Eds.). (1988). *Telephone survey methodology.* New York: John Wiley & Sons.

Groves, R. M., Cialdini, R. B., & Couper, M. P. (1992) Understanding the decision to participate in a survey. *Public Opinion Quarterly, 56*(4), 475–495.

Groves, R. M., & Couper, M. P. (1998). *Nonresponse in household interview surveys.* Somerset, NJ: John Wiley & Sons.

Groves, R. M., Dillman, D. A., Eltinge, J. L., & Little, R. J. A. (Eds.). (2002). *Survey nonresponse.* New York: John Wiley & Sons.

Groves, R. M., Fowler, F. J., Couper, M. P., Lepkowski, J. M., Singer, E., & Tourangeau, R. (Eds.). (2004). *Survey methodology.* New York: John Wiley & Sons.

Groves, R. M., & Kahn, R. L. (1979). *Surveys by telephone: A national comparison with personal interviews.* Orlando, FL: Academic Press.

Guensel, P. J., Berckmans, T. R., & Cannell, C. F. (1983). *General interviewing techniques.* Ann Arbor, MI: Survey Research Center.

Hatchett, S., & Schuman, H. (1975–1976). White respondents and race-of-interviewer effects. *Public Opinion Quarterly, 39,* 523–528.

Hidiroglou, M. A., Drew, J. D., & Gray, G. B. (1993). A framework for measuring and reducing nonresponse in surveys. *Survey Methodology, 19*(1), 81–94.

Hippler, H. J., Schwarz, N., & Sudman, S. (1987). *Social information processing and survey methodology.* New York: Springer-Verlag.

Hochstim, J. R. (1967). A critical comparison of three strategies of collecting data from households. *Journal of the American Statistical Association, 62,* 976–989.

Huggins, V., & Eyermen, J. (2001). *Probability based Internet surveys: A synopsis of early methods and survey research results.* Paper presented at the Federal Committee on Statistical Methodology Conference, Washington, DC.

Jabine, T. B., Straf, M. L., Tanur, J. M., & Tourangeau, R. (Eds.). (1984). *Cognitive aspects of survey methodology: Building a bridge between disciplines: Report of the advanced research seminar on cognitive aspects of survey methodology.* Washington, DC: National Academy Press.

James, J. M., & Bolstein, R. (1990). Monetary incentives and follow-up mailings. *Public Opinion Quarterly, 54*(3), 346–361.

Johnson, T. P., Parker, V., & Clements, C. (2001). Detection and prevention of data falsification in survey research. *Survey Research, 32*(3), 1–2.

Kalton, G. (1981). *Compensating for missing survey data.* (Report for Department of Health and Human Services Contract HEW-100–79–0127, Survey Research Center, Institute for Social Research). Ann Arbor: The University of Michigan.

Kalton, G. (1983). *Introduction to survey sampling.* Newbury Park, CA: Sage.

Kanninen, B. J., Chapman, D. J., & Hanemann, W. M. (1992). Survey data collection: Detecting and correcting for biases in responses to mail and telephone surveys. In *Conference Proceedings, Census Bureau 1992 Annual Research Conference* (pp. 507–522). Arlington, VA: U.S. Department of Commerce.

Kasprzyk, D., Duncan, G., Kalton, G., & Singh, M. P. (Eds.). (1989). *Panel surveys.* New York: John Wiley & Sons.

Keeter, S., Miller C., Kohut A., Groves R. M., & Presser S. (2000). Consequences of reducing nonresponse in a national telephone survey. *Public Opinion Quarterly, 64*(2), 125–148.

Kinsey, S. H., & Jewell, D. M. (1998). A systematic approach to instrument development in CAI. In M. Couper, R. Baker, J. Bethlehem, C. Clark, J. Martin, W. Nicholls, et al. (Eds.). *Computer assisted survey information collection* (pp. 105–123). New York: John Wiley & Sons.

Kish, L. (1965). *Survey sampling.* New York: John Wiley & Sons.

Kish, L. (1987). *Statistical design for research.* New York: John Wiley & Sons.

Kulka, R. A. (1995). *The use of incentives to survey "hard-to-reach" respondents: A brief review of empirical research and current research practice.* Paper presented at the Federal Committee on Statistical Methodology Conference: Seminar on New Directions in Statistical Methodology, Washington DC.

Lachin, J. M. (1981). Introduction to sample size determination and power analysis for clinical trials. *Controlled Clinical Trials, 2,* 93–113.

Lachin, J. M. (1998). Sample size determination. In P. Armitage & T. Colton (Eds.), *Encyclopedia of biostatistics* (Vol. 5, pp. 3892–3903). New York: John Wiley & Sons.

Lessler, J. T., & Kalsbeek, W. D. (1992). *Nonsampling error in surveys.* New York: John Wiley & Sons.

Locander, W., Sudman, S., & Bradburn, N. M. (1976). An investigation of interview method, threat, and response distortion. *Journal of the American Statistical Association, 71,* 269–75.

Lyberg, L., Collins, M., de Leeuw, E., Dippo, C., Schwarz, N., & Trewin, D. (Eds.). (1997). *Survey measurement and process quality.* Somerset, NJ: John Wiley & Sons.

Mangione, T. W. (1995). *Mail surveys: Improving the quality.* Thousand Oaks, CA: Sage Publications.

Marsden, P. V. (Ed.). (1994). *Sociological methodology* (Vol. 24). Cambridge, MA: Blackwell.

Moser, C. A., & Kalton, G. (1972). *Survey methods in social investigation* (2nd ed.). New York: Basic Books.

Neyman, J. (1934). On the two different aspects of the representative method: The method of stratified sampling and the method of purposive selection. *Journal of the Royal Statistical Society, 97,* 558–606.

Nicholls II, W. L., Baker, R. P., & Martin, J. (1997). The effect of new data collection technologies on survey data quality. In L. Lyberg, P. Biemer, M. Collins, E. de Leeuw, C. Dippo, N. Schwarz, et al. (Eds.), *Survey measurement and process quality* (pp. 221–248). New York: John Wiley & Sons.

Oldendick, R. W., Bishop, G. F., Sorenson, S. B., & Tuchfarber, A. J. (1988). A comparison of the Kish and last-birthday methods of respondent selection in telephone surveys. *Journal of Official Statistics, 4*(4), 307–318.

Oldendick, R. W., & Link, M. W. (1994). The answering machine generation: Who are they and what problem do they pose for survey research? *Public Opinion Quarterly, 58*(2), 264–273.

O'Rourke, D., & Blair, J. (1983). Improving random respondent selection in telephone surveys. *Journal of Marketing Research, 20,* 428–432.

Parsons, J., Owens, L., & Skogan,. (2002). Using advance letters in RDD surveys: Results of two experiments. *Survey Research, 33*(1), 1–2.

Payne, S. L. (1951). *The art of asking questions.* Princeton, NJ: Princeton University Press.

Piekarski, L. (1997). Unlisted telephone rates in 320 metropolitan statistical areas. *The Frame.* Survey Sampling International.

Presser, S., & Blair, J. (1994). Survey pretesting: Do different methods produce different results? *Sociological Methodology, 24,* 73–104.

Presser, S., Rothgeb, J. M., Couper, M. P., Lessler, J. T., Martin, E., Martin, J., et al. (Eds.). (2004). *Methods for testing and evaluating survey questionnaires.* New York: John Wiley & Sons.

Rasinski, K. A., Mingay, D., & Bradburn, N. M. (1994). Do respondents really mark all that apply? *Public Opinion Quarterly 58*(3), 400–408.

Rizzo, L., Brick, J. M., & Park, I. (2004). A minimally intrusive method for sampling persons in random digit dial surveys. *Public Opinion Quarterly, 68*(2), 267–274.

Rosner, B. A. (2000). *Fundamentals of biostatistics* (5th ed.). Pacific Grove, CA: Duxbury.

Royston, P. N. (1989). Using intensive interviews to evaluate questions. In F. J. Fowler (Ed.), *Conference proceedings, health survey research methods* (pp. 3–8). Washington, DC: National Center for Health Services Research.

Salmon, C. T., & Nichols, J. S. (1983). The next-birthday method of respondent selection. *Public Opinion Quarterly, 47,* 270–276.

Sarndal, C.-E., Swensson, B., & Wretman, J. (1992). *Model-assisted survey sampling.* New York: Springer-Verlag.

Schaefer, D., & Dillman, D. A. (1998). Development of a standard e-mail methodology: Results of an experiment. *Public Opinion Quarterly, 62*(3), 378–397.

Schechter, S., & Blair, J. (2001). Expanding cognitive laboratory methods to test self-administered questionnaires. *International Journal of Cognitive Technology, 6*(2), 26–32.

Schuman, H., & Presser, S. (1981). *Questions and answers in attitude surveys: Experiments on question form, wording, and context.* Orlando, FL: Academic Press.

Schwarz, N., Strack, F., & Mai, H.-P. (1991). Assimilation and contrast effects in part-whole question sequences: A conversational analysis. *Public Opinion Quarterly, 55*(1), 3–23.

Schwarz, N., & Sudman, S. (Eds.). (1992). *Context effects in social and psychological research.* New York: Springer-Verlag.

Schwarz, N., & Sudman, S. (Eds.). (1994). *Autobiographical memory and the validity of retrospective reports.* New York: Springer-Verlag.

Schwarz, N., & Sudman, S. (Eds.). (1996). *Answering questions: Methodology for determining cognitive and communicative processes in survey research.* San Francisco: Jossey-Bass.

Schwarz, N., & Wellens, T. (1997). Cognitive dynamics of proxy reporting: The diverging perspectives of actors and observers. *Journal of Official Statistics, 13*(2), 159–179.

Shettle, C., & Mooney, G. (1999). Monetary incentives in U. S. government surveys. *Journal of Official Statistics, 15*(2), 231–250.

Singer, E., Van Hoewyk, J., & Maher, M. P. (2000). Experiments with incentives in telephone surveys. *Public Opinion Quarterly, 64*(2), 171–188.

Sirken, M. G., Herrmann, D. J., Schechter, S., Schwarz, N., Tanur, J. M., & Tourangeau, R. (Eds.). (1999). *Cognition and survey research*. New York: John Wiley & Sons.

Strack, F. (1992). "Order effects" in survey research: Activation and information functions of preceding questions. In N. Schwarz & S. Sudman (Eds.). *Context effects in social and psychological research* (pp. 23–34). New York: Springer-Verlag.

Sudman, S. (1967). *Reducing the cost of surveys*. Chicago: Aldine.

Sudman, S. (1976). *Applied sampling*. New York: Academic Press.

Sudman, S., & Bradburn, N. M. (1982). *Asking questions: A practical guide to questionnaire design*. San Francisco: Jossey-Bass.

Sudman, S., Bradburn, N. M., & Schwarz, N. (1996). *Thinking about answers: The application of cognitive processes to survey methodology*. San Francisco: Jossey-Bass.

Tanur, J. M. (1992). *Questions about questions: Inquiries into the cognitive bases of surveys*. New York: Russell Sage Foundation.

Tourangeau, R., Couper, M. P., & Steiger, D. M. (2001). *Social presence in web surveys*. Paper presented at the Federal Committee on Statistical Methodology Conference, Washington, DC.

Tourangeau, R., & Rasinski, K. A. (1988). Cognitive processes underlying context effects in attitude measurement. *Psychological Bulletin, 103*, 299–314.

Tourangeau, R., Rips, L., & Rasinski, K. (2000). *The psychology of survey response*. New York: Cambridge University Press.

Tourangeau, R., & Smith, T. W. (1996). Asking sensitive questions: The impact of data collection mode, question format and question context. *Public Opinion Quarterly, 60*(2), 275–304.

Tripplett, T. (1994, May). *What is gained from additional call attempts and refusal conversion and what are the cost implications?* Paper presented at the International Field Directors and Field Technicians Conference, Boston, MA.

Troldahl, V. C., & Carter, Jr., R. E. (1964). Random Selection of Respondents within Households in Phone Surveys. *Journal of Marketing Research, 1*(2), 71–76.

Tucker, C., Lepkowski, J. M., & Piekarski, L. (2002). The current efficiency of list assisted telephone sampling designs. *Public Opinion Quarterly, 66*(3), 321–338.

Turner, C. F., & Martin E. (Eds.). (1984). *Surveying subjective phenomena* (Vol. 1). New York: Russell Sage Foundation.

U.S. Census Bureau. (1989). *Bicentennial census facts: 1990 Decennial census*. Washington, DC: U.S. Government Printing Office.

U.S. Census Bureau. (1993). *1990 Census of population and housing: Population and housing unit counts—North Carolina* (1990 CPH-2-35). Washington, DC: U.S. Government Printing Office.

U.S. Census Bureau. (2000a). *Profile of selected housing characteristics: 2000.* [MRDF] Quick tables—American FactFinder. Retrieved February 6, 2004 from http://factfinder.census.gov.

U.S. Census Bureau. (2000b). *Telephone service available.* [MRDF] Detail tables— American FactFinder. Retrieved February 6, 2004 from http://factfinder. census.gov.

U.S. Census Bureau. (2000c). *Tenure by poverty status in 1999 by telephone service available.* [MRDF] Detail tables—American FactFinder. Retrieved April 29, 2004 from http://factfinder.census.gov.

U.S. Census Bureau. (2000d). *Profile of general demographic characteristics: 2000.* [MRDF] Quick Tables-American FactFinder. Retrieved February 9, 2004 from http://factfinder.census.gov.

U.S. Census Bureau. (2003). *Annual demographic survey: Current population survey March supplement 2002.* Washington, DC: U.S. Dept. of Commerce (http://www.bls.ccensus.gov/cps/datamain.htm).

U.S. Census Bureau. (2004). *Meeting 21st century demographic data needs— Implementing the American community survey. Report 4: Comparing general demographic and housing characteristics with census 2000.* Washington. DC: U.S. Government Printing Office.

Waksberg, J. (1978). Sampling methods for random digit dialing. *Journal of the American Statistical Association, 73*(361), 40–46.

Warriner, K., Goyder, J., Gjertsen, H., Hohner, P., & McSpurren, K. (1996). Cash versus lotteries versus charities in mail surveys. *Public Opinion Quarterly, 60*(4), 542–562.

Willimack, D., Lyberg, L., Martin, J., Japec, L., & Whitridge, P. (2004). Evolution and adaptation of questionnaire development: evaluation and testing methods in establishment surveys. In S. Presser, J. M. Rothgeb, M. P. Couper, J. T. Lessler, E. Martin, J. Martin, et al. (Eds.). *Methods for testing and evaluating survey questionnaires.* New York: John Wiley & Sons.

Willis, G. (1994). *Cognitive interviewing and questionnaire design: A training manual.* Working paper series. Hyattsville, MD: National Center for Health Statistics.

Willis, G. B., Trunzo, D. B., & Strussman, B. J. (1992). The use of novel pretesting techniques in the development of survey questionnaires. *Proceedings of the Section on Survey Research Methods, American Statistical Association,* 824–828.

Wiseman, F. (1972). Methodological bias in public opinion surveys. *Public Opinion Quarterly, 36*(1), 105–108.

Yammarino, F. J., Skinner, S. J., & Childers, T. L. (1991). A meta-analysis of mail surveys. *Public Opinion Quarterly, 55*(4), 613–639.

Glossary/Index

About the Authors

Ronald Czaja is Associate Professor of Sociology and Anthropology at North Carolina State University. He teaches courses in both undergraduate and graduate research methodology and medical sociology. His methodological research interests focus on sampling rare populations, response effects in surveys, and the cognitive aspects of questionnaire design. Early in his career, he worked at the Survey Research Laboratory, University of Illinois at Chicago, as project coordinator, co-head of sampling, assistant director, and principal investigator.

Johnny Blair is Senior Survey Methodologist and Principal Scientist at Abt Associates Inc. He has conducted methodological research on pretesting techniques, particularly cognitive interviewing, and on sample designs for special or rare populations, among other areas. He has consulted on survey design for many public and private sector organizations. Prior to joining Abt Associates, he was with the Survey Research Center, University of Maryland, 1989 to 2001, serving as acting director from 1999 to 2001. He began working in survey research at the Survey Research Laboratory, University of Illinois, Urbana–Champaign, with a focus in sampling and operations management.